Creating Shared Understanding in Product Development Teams

Louise Møller · Christian Tollestrup

Creating Shared Understanding in Product Development Teams

How to 'Build the Beginning'

Springer

Louise Møller
Aalborg University
Østeraagade 6
Aalborg 9000
Denmark

Christian Tollestrup
Aalborg University
Østeraagade 6
Aalborg 9000
Denmark

ISBN 978-1-4471-4179-2 ISBN 978-1-4471-4180-8 (eBook)
DOI 10.1007/978-1-4471-4180-8
Springer London Heidelberg New York Dordrecht

Library of Congress Control Number: 2012939041

Printed on acid-free paper

Springer is part of Springer Science+Business Media (www.springer.com)

Foreword

This is both a timely book and a timeless book, for it touches the foundation of creative collaboration among humans. In industry, academia, government, and civil society we are now constantly confronted with the challenge of finding solutions to ever more ill-defined and complex problems. While the infrastructure of the Internet has given us access to large amounts of information, we still need to work in teams and other forms of representative groups composed of members from diverse backgrounds in order to coordinate the investigation of problems and the creation of relevant, inclusive, and sustainable solutions. The way to accomplish this feat in this new world is the subject of this book.

Humans can be described as meaning making machines. From toddlers that pick up pieces of objects in their environment in order to learn more about them by tasting them, much to their parents horror, to adults asking purpose-driven questions about theirs and others desires and intentions, humans are constant seekers and makers of meaning. Yet given the natural differences in age, location, and life experiences of the members of a team, the common words we use often carry with them different degrees of meaning. This increases the risk of interpersonal misunderstandings and consequently could lead to situations of conflict. In this groundbreaking work, Louise and Christian have given us three interrelated tools to break out of this potential conundrum. First, the use of a tangible media in the form of LEGO® blocks to complement the spoken language. Second, the provision of a facilitation guide to help cope with ambiguous situations and resolve them creatively. Third, the description of a meta-cognitive framework that enables us to better understand such situations while experiencing them.

This last point is perhaps the most important. For people familiar with Mihály Csíkszentmihályi's concept of flow, *Creating Shared Understanding in Product Development Teams* can be seen as guidebook to achieve collective flow in teams. This is because it helps meet the conditions for flow by facilitating novel and creative interactions between members of a team, and providing the meta-cognitive framework to assure them that they are capable of meeting the challenges of the task at hand. This is ultimately a book on ways to create coherence among team members and enhance their ability to discover and develop effective solutions, while making sure that the experience of working in a team is a positive one. I highly recommend it for leaders, managers, and team members seeking to enhance their collective experience and shared understanding.

Stanford University, California, 5 March 2012 Ade Mabogunje

Preface

It is the first meeting in a new promising project. Everyone around the table is enthusiastic about the project and excited to get going. There is a good atmosphere and the discussion travels back and forth. Everyone tries to present their point of view, which results in a broad discussion on very different aspects of the project.

It is discussed how the project should be understood, approached, and developed. At the end of the meeting, some decisions are made in relation to the project. Many decisions are related to the different deliverables for the next meeting. At some point someone asks if they have reached an agreement and everyone nods their approval. Everyone leaves the meeting, confident that they know what to do.

A few weeks later, it is time for the second meeting. The team spirit is still high and there is a nice buzz in the meeting room, before the meeting starts. The introduction proceeds without problems, and it is time to recap what has been done in the project since the last meeting. The different participants start presenting their promised deliverables.

In the beginning everything seems fine; however, after a few presentations it is clear that there are very different understandings of the project as well as the assignments for the meeting. In fact it seems as if the participants have been working in different directions and with different aims.

More and more questions are asked and soon the presentations turn into a discussion about understanding the deliverables, the project, and its aim.

The positive and enthusiastic atmosphere is soon taken over by mild frustration and a slight disappointment.

What happened? A few weeks ago everyone nodded their approval, and seemed confident that they knew what to do. Now, it seems as if everyone is pursuing different goals and that nobody really understands each other.

The situation described above could perhaps be taken out of several different contexts and scenarios. Most people, who have been working in teams, probably recognize it, and especially people with experiences from interdisciplinary teams can confirm that this situation is part of many projects.

Lack of shared understanding or frames is just one of the difficulties facing interdisciplinary design teams working in pre-development projects. Besides managing their different values, perspectives, and interests that make them see different things as important, they also have to figure out what their users and stakeholders find important.

In other words, the team has to frame their project around real user needs, problems or opportunities—and figure out what people really want, and at the same time come to an agreement about this framing within the team.

This is quite a challenge—both in terms of enabling the team members to express their personal framing of the project, but also in terms of making users and stakeholders communicate what kind of needs or problems they have, as well as the potential opportunities they see. And finally it is a challenge in terms of creating a shared frame within the team.

In this book, this challenge is approached from a 'designerly' perspective , based on the initial assumption that the creation of physical artifacts can help both team members, users, and stakeholders overcome the boundary of not being able to define, express, and communicate how they frame a given project or make meaning in relation to their everyday life. And that this clarity will help the creation of a shared frame.

Based on empirical evidence, it was found that not all physical artifacts were able to do so, but that a small group of physical artifacts in a special setting and with a specific set of characteristics was the physical artifacts were named Personal and Shared Experiential Concepts.

The objective of this book is to review how these physical artifacts enabled and stimulated the communication between team members, users, and stakeholders in interdisciplinary teams working on pre-development projects. And also how they enabled and supported the creation of shared frames within these teams.

Finally, this book will show how to facilitate the creation of personal and shared experiential concepts in your team.

A Thesis-Based Book

This book is a re-write of Lousie Møller's PhD thesis from 2010, now adding emphasis on how to stage the creation of Personal and Shared Experiential Concepts in a workshop setting.

From 2008 to 2010 Louise Møller was a PhD fellow at the Department of Architecture and Design at Aalborg University. During this period Louise Møller had a 6 month stay at the Center for Design Research, Stanford University that helped develop central parts of the findings in the PhD Thesis.

The findings of Personal and Shared Experiential Concepts seemed to have a lot of merit for further understanding the construction of shared frames and shared understanding in pre-development projects. At the same time the facilitation of the workshop was left unexplained and uninvestigated in the thesis, but the extensive use of the facilitated workshops throughout the PhD project has generated very hands-on practical knowledge, useful for many project managers struggling with alignment and shared understanding. Thus it seemed very appropriate to restructure the thesis into this book including the facilitation aspects, used with great success in the cases of this project.

Acknowledgments

A research project based on active research activities is not a task done by one person. The research behind this book has taken a lot of effort and strong, open cooperation with many parties. And we are very grateful for this.

First and foremost, we would like to thank Professor Larry Leifer and Dr. Ade Mabogunje from Stanford Centre for Design Research, who provided the opportunity to Louise Møller to become a 'visiting scholar' at Stanford. Their fireworks of ideas, intense presence, interest, and encouragement have been amazing and they truly are the role models for western academia. The contribution of this input and sparring on the research problems and analysis of data cannot be underestimated.

Second, the open-minded attitudes and warm welcome by the participating companies and workshop participants have played a vital role in generating the empirical data and foundation for this work. So thank you very much to the participants from: TC Electronic, Red Cross, Daimler AG, Region Northern Jutland and the "Good Elderly Life" project with Copenhagen Living Lab. In Copenhagen Living Lab, special thanks to Thomas Hammer-Jakobsen. Your interest and encouragement have been more than we could be expected, and your expectations have made us work even harder.

Third, starting out a research project looking for the right direction, the new and interesting "hole" which needs to be filled, is crucial to any PhD project. The direction and tool for generating empirical data was immensely important in the forming of the research behind this book. Therefore, special thanks go to co-supervisor Poul Kyvsgaard Hansen, Center for Industrial Production, Aalborg University.

Finally, writing a thesis and a book and doing the research takes a lot of mental attention and time. So we thank Anders, August, Anne, Jacob, and Simon for bearing with us during this period.

Contents

Chapter 1
Introduction

This chapter holds the introduction to the book. It identifies the context of the study in terms of interdisciplinary teams working in the early phases of innovation and identifies the challenges, which are facing these teams as well as present approaches toward these. It shows how this study approaches these challenges in a different way, unfolds the research questions, and outlines the structure of the book.

1.1 Managing Large Pre-development Projects

Before a pre-development project becomes a development project, uncertainty and chaos may rule and the project may not be anchored very well in the organization. At the same time, it is an opportunity to explore project directions, before all criteria are defined and decisions on directions are made.

This book is concerned with large projects or project partnerships, which are positioned in the period of time before the choice of direction and anchoring in the organization has taken place. This period of time can be very long or fairly short depending on the project—but what defines it is that no strategic plans have been made, no final goals have been set, and no concept has been developed [8]. In retrospect, the process may look very linear, but when you are in the midst of it, it may seem more chaotic, and the project may take any number of directions with very different scopes (See Fig. 1.1).

In 'Innovation Management and Engineering Design', this period of time is called The Fuzzy Front End of Innovation [14], Innovation in the making [8], or Design of Business [21].

In creative design, the activities in this period of time are called Strategic Design [11], Transformation Design [3], or Concept Design [26]. However, since this book is directed toward an interdisciplinary audience, it is considered appropriate to use the general term 'The early phases of Innovation'.

L. Møller and C. Tollestrup, *Creating Shared Understanding in Product Development Teams*, DOI: 10.1007/978-1-4471-4180-8_1, © Springer-Verlag London 2013

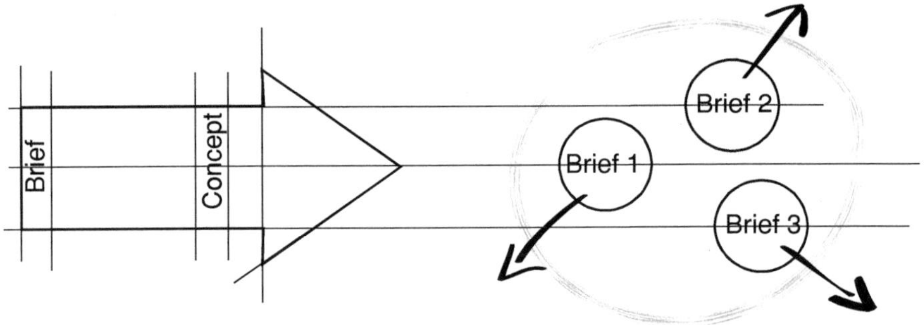

Fig. 1.1 Pre-development projects may look linear in retrospect, but signify a period where a project may take many directions

1.2 Characteristics of the Early Phases of Innovation

One of the greatest challenges in the early phases of innovation is to understand the scope of the issue and to find the right need, problem, or opportunity to approach [12]. It does not matter how well the rest of the process is managed or how well the project is accomplished, if it is based on the wrong problem or subordinate need [6]. Therefore, a great part of the effort in the early phases of innovation is used on researching and investigating the project context and interacting with users and stakeholders, in order to incorporate as well as build upon their insights and perspectives.

The early phases of innovation are further characterized by direct confrontation with abstract and strategic questions like:

> Where do we go from here? What do we want to create? For whom? And why? [16].

Accordingly, the early phases of innovation require a different way of engaging with the problem context compared to traditional problem solving. As Silje Friis describes it:

> (…) (they) do not just fix problems—they enter projects at a much earlier stage, generating new opportunities and mastering unframed problem solving ([11], p. 71).

The early phases of innovation can be organized in different ways and are of course adjusted to the nature of the specific project. However, most early phase projects include activities such as information collection/exploration [20], opportunity identification, idea generation and enrichment, opportunity analysis, idea selection, and concept definition [14].

Alternatively, the early phase of innovation has been described as a balance between exploring and framing [8] or as an iterative process of divergence and convergence [31].

In practice, it may be identified as a constant shift between exploring and unfolding the project context and combining the insights into a meaningful problem framing or design brief. In contrast to traditional design projects, a large part of the early phases is used on designing the brief:

> (…) the groups in this community of practice work 'upstream' of the traditional brief. Their involvement begins before the design brief is formulated, working with user groups and organizations to understand the scope of the issue and define the right problem to tackle ([3], p. 20).

The early phases are also characterized by an extended interaction with stakeholders. This is done to capture the different interests and perspectives in relation to the project, and to assure the stakeholders' commitment to and ownership of the project outcome [3].

The outcome of an early phase project is the problem framing in terms of a design brief or, in some cases, a number of design briefs, depending on the size of the project and the number of participants. Since design briefs come in many forms and are used in many different connections, the following definition created by Max Munnecke is found useful. The design brief consists of at least three elements:

1. A framing of the project context.
2. An understanding of the values and meanings, which users apply.
3. Insights into the dilemmas or problems, which are central to the project.

1.3 Interdisciplinary Teams

Projects in the early phases of innovation are often run and managed by teams in order to handle their size and complexity. The teams are working with open-ended and unframed objectives, and engage in different types of activities. This calls for different competences, understandings, and perspectives, and therefore the project teams are often interdisciplinary [11]. Typically, the teams include professionals like designers, managers, engineers, ethnographers, sociologists, psychologists, economists, etc.

The interdisciplinary team ensures the necessary diversity in the perspective, body of knowledge, values, and goals as well as different professional skills [26] and helps the team in their pursuit of the right need, problem, or opportunity to approach. However, the different frames and assumptions which each team member brings to the interdisciplinary team also represent a challenge [12]. It may result in misunderstandings, contradictory interpretations in worst-case conflicts, which mean that the cohesion in the team is challenged. Another possibility is that the team members end up pursuing different goals, and that the team therefore has to spend a lot of time synchronizing the different efforts.

1.4 Key Finding: Prototyping the Point of Departure

The key finding presented in this book is that the creation of physical artifacts in a specific setting, and with a specific set of characteristics can function as important drivers for communicating personal meaning making and creating shared frames in project teams. Further, it is found that these physical artifacts enable users and other stakeholders to overcome the boundary of not being able to communicate how they make meaning of their everyday life, as well as enable team members to communicate their personal framing of the project. In other words, the physical artifacts with this set of specific characteristics, reduce some of the boundaries experienced by the interdisciplinary teams, both when it comes to their internal collaboration as well as in their interaction with the users or stakeholders.

In the research process, it was further found that physical artifacts with these specific characteristics were not described in any previous research. It was therefore relevant to unfold them in this project, and give them the names Personal- and Shared Experiential Concepts. This means that Personal- and Shared Experiential Concepts are conceptualized as part of this book and represent part of the 'new knowledge' created in this research project.

Accordingly, the claims of this book is that Personal- and Shared Experiential Concepts are important drivers for communicating individual meaning and creating shared frames in interdisciplinary teams, working in the early phases of innovation. And that, Personal Experiential Concepts enable users and other stakeholders to overcome the boundary of not being able to define, express, and communicate how they make meaning of their everyday life.

The vehicle for examining and unfolding these claims is a selection of video documented workshops in real time, interdisciplinary team projects with organizations including TC Electronic, Red Cross, Daimler AG, and Region Northern Jutland. Besides these, a longitudinal study of the user-driven innovation project: 'The Good Elderly Life' has also been made, which included two sequential workshops and a more in-depth analysis of the project progress.

1.5 Focus of the Book: Creating Shared Frames in Teams and Involving Stakeholders

The early phases of innovation represent the larger boundaries of this book along with large projects or project partnerships, in which interdisciplinary teams can be found. In smaller projects, it is often not possible to find interdisciplinary teams, and they are therefore outside the scope of this project.

Inside this boundary, the book focuses on two situations in particular, the first being the creation of a shared project framing of the interdisciplinary team, and the second being the interaction with stakeholders and users of the interdisciplinary team, in order to understand their insights and perspectives regarding the design context as well as the problem framing.

The first situation is the internal challenges in the interdisciplinary teams, as they try to understand the design context and create a shared problem framing. The second is the external challenges—i.e., the interdisciplinary team's analysis, observation, and interaction with stakeholders and users, in order to understand

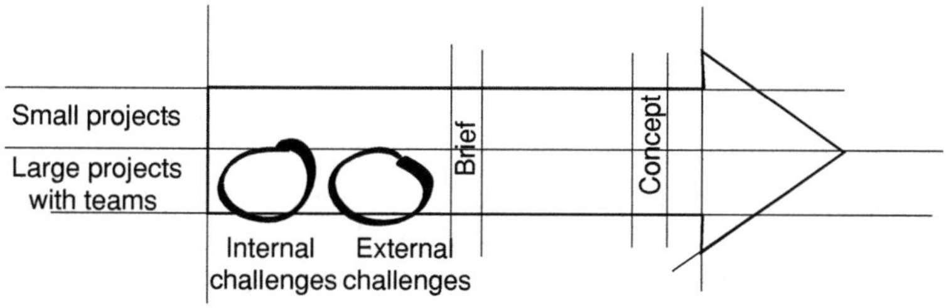

Fig. 1.2 The boundaries and focus points in this book

their insights and perspectives regarding the design context as well as the problem framing. This is also illustrated in Fig. 1.2. In the rest of the introduction, the intention is to zoom in on the two focus areas shown in Fig. 1.2.

First, the review will examine the internal and then the external challenges facing the interdisciplinary team. This will be followed by a short review of the present approaches toward these, along with a description of the gap in the present knowledge, which this book approaches.

1.6 The Challenges

Based on the literature review, it is found that there are at least three types of challenges, which interdisciplinary teams working in the early phases of innovation have to overcome. These challenges are: (1) *Diversity,* (2) *Complexity and Ambiguity, and* (3) *Asymmetry or Stickiness of information.* In the section below, a brief review of each of these follows:

1.6.1 Diversity

When a newly formed interdisciplinary team initiates an early phase project, each of its members will have an already existing set of perspectives, values, and assumptions about what is important in the given project as well as different understandings, which will direct how they think the project should be addressed [12]. Their different backgrounds, bodies of knowledge, approaches, and perspectives are enabling them to approach this project. However, along with these comes the possibility of misunderstandings, conflict, and disagreements, which can slow down the process, or even have a negative impact on performance [1, 24]. As Sessa and Jackson [28] have observed:

> Although research and theory (...) suggest that diversity have a positive impact on performance, diversity is hypothesized to have the opposite effect on cohesion' (p.140).

1.6.2 Complexity and Ambiguity

Interdisciplinary teams working in the early phases of innovation are not only challenged by their internal diversity; they also have to manage the fact that the information available in the early phases is not always complete. As mentioned earlier, one of the main objectives in the early phases of innovation is to find the right problem, need, or opportunity to approach and to identify the scope of the issue. This requires a wide divergence compared to later parts of the innovation process, and hence there will be a large set of incomplete and contradictory information, which has to be distilled into a brief. Therefore, this process can lead to both ambiguity and uncertainty within the team and potentially reduce the cohesion of the team [10].

The challenges in the early phases of innovation can be categorized as complex, because of its many 'unknown unknowns'. The team does not know the unknown information, which needs to be found, and the process/decisions only make sense in retrospect. This also means that modes of actions are limited to probing, sensing, and responding [29].

1.6.3 Asymmetry or Stickiness of Information

As already mentioned a few times, interdisciplinary teams are not only challenged internally, they are also faced with great external challenges in terms of understanding the scope of the issue and finding the right need, problem, or opportunity to approach [12].

One of the main external challenges, which the team working in the early phases of design and innovation has to approach, is the asymmetry or stickiness of information. As Von Hippel [32] argues:

> (…) each innovator will tend to develop innovations that draw on the sticky information it already has, because that is the cheapest course of action (…) this means that users as a class will tend to develop innovations that draw heavily on their own information about need and context of use. Similarly, manufacturers as a class will tend to develop innovations that draw heavily on the types of solution information in which they specialize. (p. 70).

Accordingly, the design team must be careful not to focus only on its solution capacity, but also on gaining the information about needs and context of use, which enables them to find the right problem, need, or opportunity to approach [12]. However, this is not an easy task. For many users or stakeholders, it is simply impossible to explain how they apply meaning to a situation or activity, either because they have done it so many times that they do not think about it anymore, or because they find it hard to define or articulate [15].

1.6.4 Delimitation

Apart from the challenges named above, several organizational or political challenges are also present, which might have an impact on teams working in the early phases of innovation, such as organizational resistance when it comes to implementing radical new solutions, or difficulties when initiating collaboration between different departments and so forth. But since the focus of this dissertation is on the interdisciplinary team, the organizational and political challenges are not within the scope of the research and will therefore not be described further.

Another set of challenges, which also falls outside the scope of this dissertation, are the challenges related to underlying emotions [27] and interpersonal dynamics [2].

1.7 Approaches Toward the Challenges

This section will provide an overview of the present approaches and recommendations in relation to the challenges in the early phases of innovation based on previous research within the respective fields.

1.7.1 Diversity

A theme, which seems to reappear in relation to handling diversity in teams, is *'sharing'*; especially, shared task commitment has been praised as an essential driver for collaboration and cohesion in teams. In a study from 2000, Carless and Depaola [4] found that task cohesion was a much stronger indicator for performance, compared to both social cohesion and individual attraction to the team. In other words, if team members have a shared task, they often feel more united compared to teams bound by friendship or personal interest.

Clarity of and commitment to a shared purpose or goal is another factor, which is mentioned as essential to successful team interaction [13]. Takeuchi and Nonaka [30] further observed the importance of creating a shared goal within the team itself, without interference of outside objectives and agendas.

Cohen and Bailey [5] also point to the positive association between team cohesion and team performance, implying that if the team is performing well, it is more likely to remain connected. Lipman-Blumen and Leavitt [19] further highlight the creation of a shared vision or mission to be important to the team cohesion, especially if all team members find it both vital for the project and personally ennobling.

1.7.2 Complexity and Ambiguity

'Sharing' and *'collective'* are also some of the keywords, which can be found in the recommendations on how to approach complexity and ambiguity. For instance, Darsø [8] emphasized the need for *'shared uncertainty'* in the early phases of innovation, due to the complexity of the problem. And in her study of management teams, Eisenhardt [9] observed that effective teams handled the challenge of uncertainty and incomplete information by building a *'collective intuition'*. This is supported by Leonard and Swap [17], who further acknowledge and discuss the management challenges, in terms of diffusing interpersonal conflicts, protecting challenging insights and perspectives, and the fostering of the necessary divergence without losing any perspectives.

1.7.3 Asymmetry or Stickiness of Information

The challenges in relation to asymmetry or stickiness of information are often approached by some kind of investigations and explorations of the market, the potential users or the context of use, as a basis for the problem scoping and the brief [31, 15, 18].

In the Innovation Management and Product Development communities, this kind of research is known as market research or generative research [22]. In the design community, these investigations carry names like design research, user research, or need finding [16, 22].

In general, research in the early phases of innovation can be divided into two broad categories—quantitative research and qualitative research [22].

Quantitative research represents a macro level analysis and is for instance able to identify political, environmental, social, technological, economic, and demographic changes. Quantitative market research can further provide information about market size, costumer profile, market gaps, as well as forecast trends in the marketplace. The quantitative research is useful, when it comes to understanding and identifying trends and changes in the market. It can provide an insight into *what* is going on and perhaps *how* things are changing, but unfortunately it is very limited, when it comes to understanding *why* changes occur [7].

Qualitative research, on the other hand, is useful when it comes to identifying *why* changes occur, as it is able to provide micro level contextual insights into, for instance user context, lifestyles, behavior, and values. Qualitative research is a tool for understanding user behavior, as well as for discovering unmet and unarticulated user needs [7].

Another difference between the quantitative and qualitative research approach is that the former investigates a broad and overall sample, searching for explicit and quantifiable information, whereas the latter investigates a very narrow sample, searching for tacit and emerging insights [23]. Even if the quantitative approach and qualitative approach are quite different, most researchers agree that they must be seen

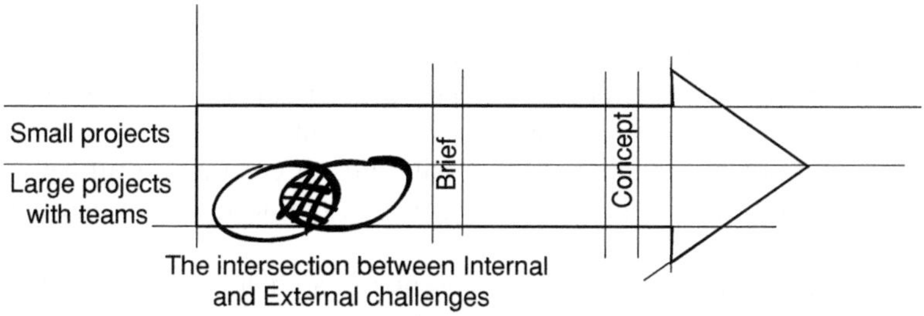

Fig. 1.3 Gap of knowledge to be investigated in this book

as complementary and that they play different roles [23, 22]. Some research groups even emphasize that in order to understand the increasing complexity of people's lives, they have to be equally valued and integrated in the innovation process [25, 7].

1.8 Gaps in the Present Knowledge

One of the main gaps in the present knowledge, which occurs when viewing the challenges in the early phases of innovation along with recommended approaches toward these, is the intersection between the internal and the external challenges.

In current research, the focus is either on the team and how to handle the diversity, complexity, or ambiguity within the team, or the focus is on the interaction between users and stakeholders and how to manage the asymmetry or stickiness of information. However, no one looks at the intersection between the two.

In practice, this division between the internal and the external challenges does not exist, as both sets of challenges are present at all times in the early phases of innovation, even if the team designates different people or different kinds of activities to approach these. The intersection between the internal and external challenges is therefore identified as a gap in the present knowledge, which will be approached in this book. This is also illustrated in Fig. 1.3.

1.9 The Research Setting and the Research Questions

The aim of this research is to build upon the present approaches or recommendations in relation to the internal and the external challenges and to look at how insights from different perspectives, approaches and methods can be integrated and synthesized. Besides this, the objective is also to introduce a more hands-on approach to the challenges.

This study will view the interdisciplinary team's work in the early phases of innovation as an effort of meaning making and thereby as a process, in which participants are translating personal insights and integrating these with user research into a comprehensive and shared project framing.

Another central issue of this research is direct interaction with users and stakeholders: meeting them in person and not only through a user research report, which can become hard to translate into products and/or services. The underlying assumption is that direct contact encourages different perspectives, expertise, and approaches and prevents attempts and possibilities of creating a wall between the designers and the researchers as well as between the design team and the users/ stakeholders.

A third and defining issue of the present research has been the creation of physical artifacts as a way of creating a shared language between the different perspectives, as well as a way to emphasize the importance of co-creation and ownership. But also to help users and stakeholders to overcome the boundary of not being able to express, what they find meaningful, and how they make meaning of their everyday activities.

The inspiration for this derives from the later parts of both design and innovation processes. Here, prototypes and models are often used to support communication, to test possibilities and compare ideas, etc.

The intention in this book has been to use the creation of physical artifacts in the very early phases, and to test their usability here, as well. However, in the early phases of innovation it is not possible to model the product or the service yet, and therefore the creation of physical artifacts is used to communicate personal meaning making in relation to either the scope of the project or the problem to be approached, depending on the time and the objective of the creation.

The creation of physical artifacts in the early phases of innovation also entails some extra demands of the *'creation process'* and facilitation of this:

The requirement for creating the physical artifacts has to be kept to a minimum in order to involve as many relevant users and stakeholders as possible (since it is not given that everyone in the interdisciplinary team or among users/stakeholders has modeling experience).

The creation of the physical artifacts has to handle a great diversity in terms of different backgrounds and assumptions as well as types and levels of knowledge. And it has to contain both the complexity and ambiguity, which may appear in relation to the project.

And finally, the creation of the physical artifacts has to involve a shared experience and shared output in order to build on the insights from previous research, in which 'sharing' is identified as an important tool to handle diversity, complexity, and ambiguity in teams.

In practice, the research setting is a number of facilitated workshops, in which the creation of physical artifacts is a central element. The workshops take place in early phase innovation projects and include both members of the interdisciplinary team as well as users and stakeholders with the relevant contextual knowledge. The aim of the workshops is to create a shared framing of the project within the interdisciplinary team and find the right problem, need, or opportunity to approach. Depending on the workshop setup, the workshop either leads to focus, priorities, or the acknowledgment that there is a need for further information. In Chap. 3, there will be a more elaborate description of the workshop setup.

The research questions in this book are also framed within this specific workshop setting, and the answers are to be understood in this context.

Furthermore, the book builds on the a priori assumption that all interdisciplinary teams will benefit from a shared frame as early as possible, even though it may be changed to another shared frame later on.

The Research Questions are:

1. How can the creation of physical artifacts enable and stimulate the communication between team members, users, and stakeholders in interdisciplinary teams working in the early phases of innovation?
2. How can the creation of physical artifacts enable and support the creation of shared frames within interdisciplinary teams working in the early phases of innovation?

And the subsequent question is then:

> How can one facilitate the situation, where these physical artifacts are used to enable and stimulate communication as well as create shared frames?

This last question leads to a number of recommendations and guides that can be found in Chap. 8: Facilitation Guide.

1.10 Summary

In this chapter, the framing of this book has been presented. The study is concerned with interdisciplinary teams working in early phases of innovation and focuses in particular on two situations. The first is the interdisciplinary team's creation of a shared project framing, and the second is the interdisciplinary team's interaction with stakeholders and users, in order to understand their insights and perspectives regarding the context as well as the problem framing.

The internal challenges (diversity, complexity, ambiguity) and the external challenges (asymmetry or stickiness of information knowledge) facing the interdisciplinary teams were reviewed along with the present recommendations in relation to these. It was found that there is a lack of research in the intersection between the internal and the external challenges, and this was therefore selected as the focus of this book. The chapter concluded with a review of the research setting and a presentation of the research questions.

1.11 Overview of Chapters

To provide a further overview of the book, Table 1.1 provides a brief summary of each chapter and their main conclusions.

Table 1.1 Summary of chapters

Chapter 1 Introduction	Chapter 1 reviews the motivation and claims of the study. It defines the early phases of innovation as the outer boundary of the study and identifies two focus points—in terms of the internal challenges and the external challenges. These challenges are unfolded and the present approaches toward these are presented. A gap in the present knowledge is identified, and the chapter is concluded with a review of the research setting and the research questions.
Chapter 2 Theoretical Framework	Chapter 2 holds the theoretical foundation. It includes both a review of the design perspective used in this book, as well as a presentation of the theoretical framework. The review of the design perspective includes different insights on design—and concludes with the argumentation in respect to why the design perspective is relevant in relation to the early phases of innovation. In the theoretical framework, meaning is reviewed in relation to individuals, teams, and artifacts. This review includes insights in relation to individual meaning making, the creation of second-order understandings, individual and shared frames, metaphors, models, prototypes, boundary objects, etc.
Chapter 3 Workshop Cases and Method	Chapter 3 holds a review of the workshop method and the research material. First, the workshop method: Lego Serious Play is reviewed, along with the adjustments that have been made to make it fit projects in the early phases of innovation. Second, the research material is presented. The presentation shows the workshop portfolio, and the broad variety of projects and organizational contexts, where the data come from, as well as its real time qualities. After this, each workshop is presented in terms of context, type, assignment, and participants. And finally, a large set of the physical artifacts from the workshop is presented in terms of pictures and transcripts.
Chapter 4 Identifying General Patterns	In Chapter 4, the general patterns across the six workshops are presented. It is found that some of the models played a more significant role in the workshops compared to other models. And that these models influenced both the communication of meaning and the creation of shared frames in the workshops. When analyzing the significant models closely, it is further found that their structure was different from the other models, as they had both a 'concept component' and an 'experience component'. The number of significant models across the workshops is also presented along with several examples from the workshops. In the last part of the chapter, the focus is on reviewing the transformation of the significant models from individual models to shared models.

(continued)

Table 1.1 (continued)

Chapter 5 The Significant Models	In Chapter 5, the general findings in relation to the significant models are unfolded. It is also reviewed why the significant models influence the communication of sticky information and meaning making the way they do, and why they support the creation of shared frames in the teams. In relation to this, the structure of the significant models (the 'concept component' and the 'experience component') is identified as very influential.
Chapter 6 Personal and Shared Experiential Concepts	In Chapter 6, the significant models are explored in terms of definitions. Initially, both boundary objects and metaphors seem to be plausible definitions. However, it is found that none of these definitions fully covers the characteristics of the significant models. It is therefore found necessary to understand the significant models as something new and not previously defined. At the end of the chapter, the characteristics of the significant models are gathered on the basis of the insights from the previous chapters. And the significant models are given the names Personal-and Shared Experiential Concepts.
Chapter 7 Conclusion	Chapter 7 holds the conclusion of the book as well as a synthesis of the different theoretical and empirical parts of the research project. First, it completes the research circle by answering the research questions. Second, it unfolds the reliability of the research project and discusses the conclusions with respect to the research design. Third, the theoretical foundation is discussed in respect to the findings, and finally the perspectives and further implications from this study are reviewed and discussed.
Chapter 8 Facilitation Guide	Chapter 8 holds a facilitation guide, which shows how to plan and execute a team and stakeholder workshop in the early phases of innovation. The chapter provides recommendations for the practical planning of a workshop as well as guidelines for the facilitator's role and mindset in the workshop. Furthermore, it reviews and unfolds a facilitation plan from one of the workshops presented in this book, which can be used as the starting point for developing and facilitating a team or stakeholder workshop in the early phases of innovation.

References

1. Ancona DG, Caldwell DF (1992) Demography and design: predictors of new product team performance. Organ Sci 3(3):321–341
2. Bradford DL, Huckabay MA (1999) The interpersonal dynamics reader graduate school of business. Stanford University, Palo Alto
3. Burns C, Cottam H, Vanstone C and Winhall J (2006). Transformation design council, UK available at http://www.designcouncil.info/mt/RED/transformationdesign/TransformationDesignFinalDraft.pdf

4. Carless SA, DePaola C (2000) The measurement of cohesion in work teams. Small Group Res 31(1):71–88
5. Cohen SG, Bailey DE (1997) What makes teams work: group effectiveness research from the shop floor to the executive suite. J Manag 23(3):239–290
6. Cooper RG (1998) Predevelopment activities determine new product success Industrial Marketing Management vol. 17, pp 237–247
7. Cooper R, Evans M (2006) Breaking from tradition: market research consumer needs, and design futures. Des Manag Rev 17:68–76
8. Darsø L (2001) Innovation in the making. Samfundslitteratur, Copenhagen
9. Eisenhardt KM (1999) Strategy as strategic decision making. Sloan Manag Rev 40(3):65
10. Eisenhardt KM, Tabrizi BN (1995) Accelerating adaptive processes: product innovation in the global computer industry. Adm Sci Q 40:84–110
11. Friis SK (2007) Conscious design practice as a strategic tool. Dissertation, Danish School of Education, Copenhagen
12. Hey JHG, Joyce CK, Beckman SL (2007) Framing innovation: negotiating shared frames during early design phases. J Des Res 6(1–2):79–99
13. Katzenbach JR, Smith DK (1993) The wisdom of teams: creating the high performance organization. Harvard Business School Press, Boston
14. Koen PA, Ajamian GM, Boyce S, Clamen A, Fisher E, Fountoulakis S, Johnson A, Puri P (2002) Fuzzy front end: effective methods, tools and techniques, in Belliveau P (ed.) The PDMA Tool book for new product development. Wiley, London
15. Krippendorff K (2006) The semantic turn—a new foundation for design CRC Press/Taylor and Francis Group, New York
16. Laurel B (2003) Design research methods and perspectives The MIT press, London
17. Leonard DA, Swap WC (1999) When sparks fly. Igniting Creativity in Groups Harvard Business School Press, Boston
18. Lindsay C, Rocchi S (2004) Context-of-use co-research methodology. In: Manzini E, Collina L, Evans S (eds) Solution oriented partner ship–how to design industrialized sustainable solutions. Cranfield University, Cranfield
19. Lipman-Blumen J and Leavitt HJ (1999) Hot groups: seeding them, feeding them, and using them to ignite your organization Oxford University Press, New York
20. March JG (1991) Exploration and exploitation in organizational learning. Organ Sci 2(1):71–87
21. Martin R (2004) The design of business. Rotman management (the alumni magazine), winter 2004, pp 7–10
22. Merholz P, Wilkens T, Schauer B and Verba D (2008) Subject to change: great products and services for an uncertain world O'Reilly Media, Sebastopol
23. Munnecke M and Van der Lugt R (2006) Bottom-up strategies in consumer-led markets. Second international Seville seminar on Future-Orient ed Technology analysis: Impact if FTA approaches on policy and decision making. Seville, September 28–29
24. Pelled LH, Eisenhardt KM, Xin KR (1999) Exploring the black box: an analysis of work group diversity, conflict and performance. Adm quart 44(1):1–28
25. Rocchi S, Lindsay C (2004) User in context-of-use. In: Manzini E, Collina L, Evans S (eds) Solution oriented partnership—how to design industrialized sustainable solutions. Cranfield University, Cranfield
26. Rosted J, Lau T, Høgenhaven C and Johansen P (2007) Concept design—how to solve complex challenges of our time FORA, Copenhagen. available at http://www.ebst.dk/file/7661/conceptdesign.pdf)
27. Schein EH (1999) What to observe in a group, in N.T.L. Reading book for human relation training 8th edn. N.T.L Institute of applied behavioral science VA, pp 267–269
28. Sessa VI and Jackson SE (1995) Diversity in decision making teams: all differences are not created equal, in Chemers MM, Oskamp S, Costanzo MA (eds.) Diversity in organizations. New perspectives for a changing workplace Sage Publications, Thousand Oaks

29. Snowden DJ and Boone ME (2007) A leaders framework for decision making—wise executives tailor their approach to fit the complexity of the circumstances they face Harvard Business Review (November), p 2
30. Takeuchi H and Nonaka I (1986) The new product development game. Harvard Business Review. (Jan–Feb.), pp 137–146
31. Ulrich KT, Eppinger SD (2005) Product design and development. Mcgraw-Hill Book Co, New York
32. Von Hippel E (2005) Democratizing Innovation. The MIT Press, Cambridge

Chapter 2
Theoretical Framework

In this chapter the theoretical foundation will be unfolded. A part of the theoretical foundation has already been unfolded in the introduction chapter. This was done by identifying the intersection between the internal and external challenges, which face interdisciplinary teams working in the early phases of innovation, as a gap in the present knowledge (See Fig 1.3). Still, it is necessary to unfold the theoretical foundation in two additional directions. First of all, it is necessary to provide an understanding of the perspective, which is used in this book. As mentioned earlier, the book is directed toward an interdisciplinary audience; it is, however, built upon a design perspective and a 'designerly' way of understanding the early phases of innovation. This design perspective influences for instance the development of the workshop setup. Therefore, it is necessary to unfold these implicit insights and understandings from design in this chapter and also to explain, why the design perspective is relevant in respect to the early phases of innovation. The second direction is the theoretical framework. The theoretical framework is identified and developed in order to understand and interpret the empirical findings. The theoretical framework is going to be used as the lens in which the empirical data (and the complexity it holds) can be framed and understood. The theoretical framework will have an overall focus on meaning. Meaning will be reviewed in relation to three areas, which are relevant to the workshop setup. First of all, meaning will be reviewed in relation to how individuals are creating meaning, and how this meaning is communicated. Second, meaning will be reviewed in relation to how teams are creating shared frames. And finally, meaning will be unfolded in relation to physical artifacts in terms of how they impact the communication of meaning as well as the creation of shared frames. An overview of the theoretical foundation is shown in Fig. 2.1.

L. Møller and C. Tollestrup, *Creating Shared Understanding in Product Development Teams*, DOI: 10.1007/978-1-4471-4180-8_2, © Springer-Verlag London 2013

2.1 The Design Perspective

In the following section the design perspective is presented. In detail, the objective is to present insights and understandings in relation to the following questions:

- How are problems or assignments understood and approached in design?
- What is the center of attention or value criteria in design?
- How can the process of designing be understood?
- In which ways is the design perspective relevant and useful in relation to the early phases of innovation?

The section is not to be seen as a comprehensive review in relation to the questions, but rather as a brief glimpse into the field. Likewise, only the insights and perspectives considered to be important to this book are presented.

2.1.1 Problems and Assignments in Design

Even though design is recognized as a problem-solving activity [37], it is argued that problem solving in a design context cannot be understood as it is in a mathematical/analytical context [11, 22, 31]. Instead it is argued that designers use design thinking and not analytical thinking, when solving problems.

According to Roger Martin [22] analytical thinking is characterized by a knowledge funnel, which gradually takes the mystery of every phenomenon and translates it into heuristics. The heuristics are later transformed into an algorithm and finally—what once was a mystery—may be translated into some sort of binary code. This is also illustrated in Fig. 2.2.

Design thinking, on the other hand, is more concerned with the reformulation of the mystery that is escaping the present realm of logic (or generally accepted knowledge funnel) and creating the basis for a new one.

In design methodology, this process of reformulating the 'mystery' is often referred to as repositioning [4] or reframing. Buchanan provides the following illustrative example of repositioning:

> Traditional graphic design yielded larger signs, but no apparent improvement in navigation—the larger the sign, the more likely people were to ignore it. Finally, a design consultant suggested that the problem should be studied from the perspective of the flow of the costumer experience. After a period of observing shoppers walking through stores, the consultant concluded that people often navigate among different sections of a store by looking for the most familiar and representative examples of a particular type of product. This led to a change in display strategy, placing the products that people are most likely to identify in prominent positions ([4], p. 12).

The reformulation of the mystery or escape from the present realm of logic is also represented in the kind of problems facing designers, and the way in which designers approach problems, in general.

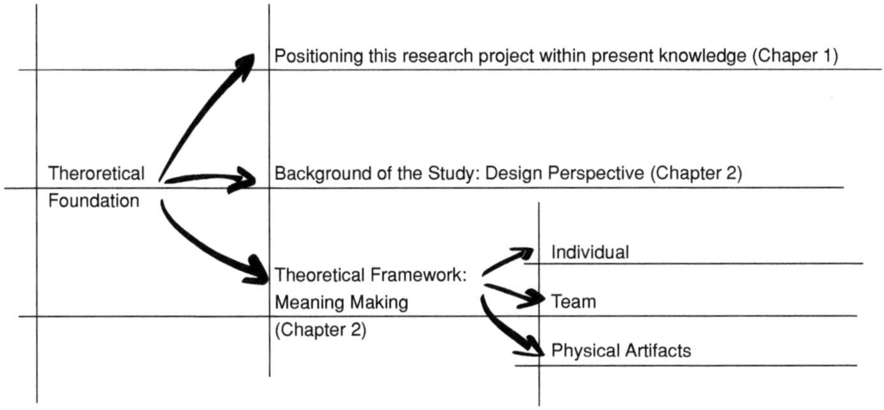

Fig. 2.1 Overview of the theoretical foundation

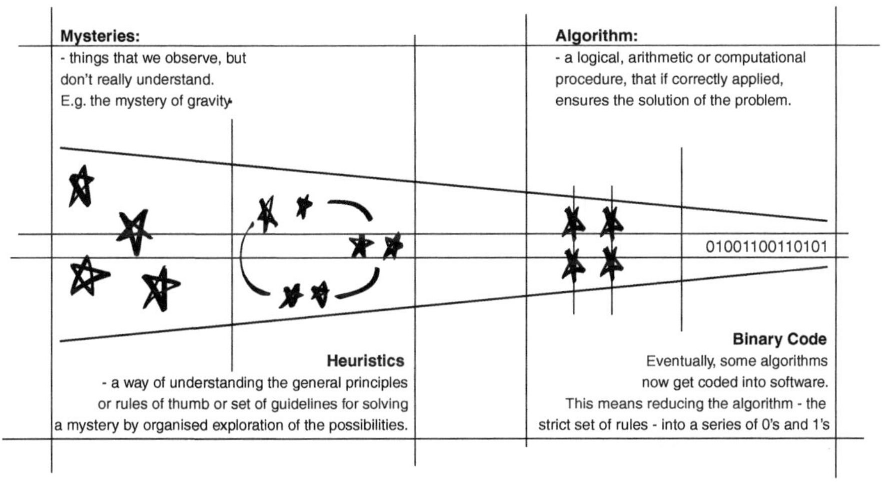

Fig. 2.2 The Knowledge funnel (Based on Martin's own illustration presented at the CONNECTING 07 conference in San Francisco, autumn 2007)

In 1972, Rittel introduced the term 'wicked problems' as a means to understand the types of problems designers are facing. Rittel [27] argues that in contrast to traditional analytical problem solving, with a clear definition and one solution, designers are faced with problems with no clear definition and therefore multiple possible solutions. In opposition to the wicked problems, he also introduced the term 'tame problems' and explained their difference as follows:

> (…) tame problems can be exhaustively formulated so that it can be written down on a piece of paper which can be handed to a knowledgeable man who will eventually solve the problem without needing any additional information. This is not so with wicked problems. When I tell somebody the problem is (…) to introduce a new product into our production

> line, I can write it down on a piece of paper, give it to him and lock him op. But it will not be long before this person will come out again and ask for more information: What kind of a new product are you talking about? How will it affect the other products already in operation? What markets do you expect for your product? etc. (…) ([27], p. 392).

Based on Rittel's work, Buchanan [4] has listed the attributes of wicked problems:

- Wicked problems have no definitive formulations, but every formulation of a wicked problem corresponds to the formulation of a solution.
- Wicked problems have no stopping rules.
- Solutions to wicked problems cannot be true or false only good or bad.
- In solving wicked problems there is no exhaustive list of admissible operations.
- For every wicked problem there is always more than one possible explanation, with explanations depending on the 'Weltanschauung' of the designer.
- Every wicked problem is a symptom of another, "higher-level" problem.
- No formulation and solution of a wicked problem has a definitive test.
- Solving a wicked problem is a "one shot" operation with no room for trial and error [after the implementation/commercialization].
- Every wicked problem is unique.
- The wicked problem solver has no right to be wrong; they are fully responsible for their actions ([4], p. 16).

The generally accepted understanding that designers are solving wicked problems also indicates that designers have a special approach in relation to the problem framings, problem scopes and problem formulations. According to Thomas and Carroll [41]:

> Design is a type of problem solving in which the problem solver views the problem or acts as though there is some ill-definedness in the goals, initial conditions or allowable transformations (p. 5).

And as a result of this, the designer will approach all problems as though they are ill-defined—regardless of whether they are or not [41]. This means that when the designer is given a problem formulation, he or she will look at the problem as only loosely 'defined' and assume that the project goal will be redefined during the project. Or as Jones [16] argues, the design brief will be seen as a kind of map for an unknown territory, rather than a specification for the solution.

2.1.2 Center of Attention or Value Criteria in Design

Another characteristic of design, which is described in many different ways, is its human-centeredness. This implies that designing is characterized by great attention and commitment in relation to the user or community for whom the design is intended. Therefore, the value of a design is constantly compared to the value it may or may not have to its users. Design activities are even argued to distinguish themselves from other creative and purposeful activities by their human-centeredness [17].

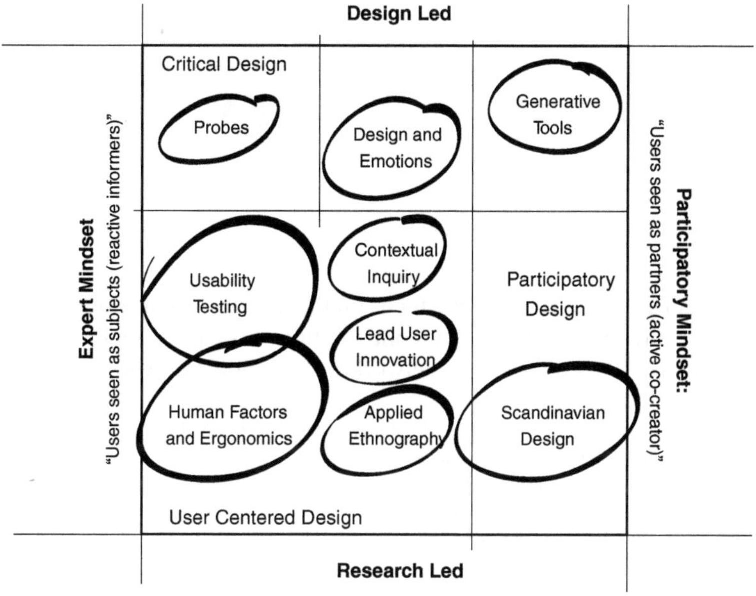

Fig. 2.3 Topography of User Research [30], own illustration

> Designers' extraordinary sensitivity to what artifacts mean to others, users, bystanders, critics, if not to whole cultures, has always been an important but rarely explicit acknowledged competence ([17], p. 48).

The human-centeredness of design is also evident in the myriad of methods in design also called design-research, user-research, or need finding [19, 23]. In 2006, Liz Sanders created Topography of User Research, in which she placed all the present methods and tools in a framework defined by the origin of the method and the mindset applied, when using it. Instead of ranging the different methods, her intention was to create a mental picture of the present approaches and mindsets used in user-research and to present these as equally valuable ways of approaching the challenges of human-centeredness. The topography can be seen in Fig. 2.3.

However, as Sanders' Topography illustrates, there is still a very lively discussion in the design community about how to view the user. The understanding of the user goes from a position, where the user is seen as a subject (reactive informer) to a position, where the user is seen as a partner (active co-creator). From a position where:

> Researchers talk about the people that they do research on as subjects, or informers or users. The people are asked questions and/or requested to respond to certain stimuli and/or observed (…) ([30], p. 5).

To a position where the designers invite the group of people, who the design is intended to benefit, to take part in the process as co-creators or partners instead of perceiving them as subjects.

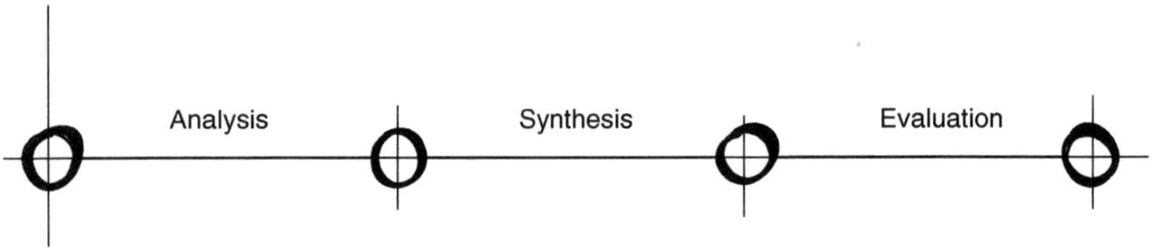

Fig. 2.4 The most basic design process model

Krippendorff [17] argues that the focus on the user in design is often misguided, and does not involve the necessary collaboration. Instead of looking at it as the user he suggests looking at it as a network of stakeholders, who:

- Are experts in their own worlds and usually are very knowledgeable about the stakes they claim in (a certain) development.
- Are willing to act in support or opposition of (a) development.
- Are willing to mobilize the resources they command: information, expertise, money, time, connections to members of their community, and the power of the institutional roles they occupy ([17], p. 64–65).

This is supported by Bucciarelli [3], who described designing in design teams as a process of achieving consensus among a group of participants with different interests. According to Bucciarelli, this process is necessarily social and requires participants to negotiate their different perspectives and construct meaning through direct interactions. However, Bucciarelli is not the only one, who has tried to explain the process of design. Like the myriad of methods to user-research and need finding there is also quite an extensive catalog of design process models.

2.1.3 The Process of Designing

In the design literature, various examples of design process models can be found, which assume that design can be organized in an identifiable process [42, 43]. Especially, the first generation models, which were introduced in the beginning of the 1960s, assumed that the design process could be divided into a set of discrete steps, which—when followed—would result in a design [28]. Most first generation models are based on the model shown in Fig. 2.4.

The underlying drive in the first generation models was to produce an approach to design based on objectivity and rationality—an approach to design in accordance with the values in science. Within this was also the wish to move from individual, intuitive, and experience based approaches to design into more stringent and explicit approaches.

Ever since the introduction of the first generation models (and later second and third generation models) a vivid discussion in the design community has taken place regarding to which extent the models are useful, and to which extent they can be ignored. The argument to keep the design process models is that the increased complexity in the design projects as well as the need to work together with other professionals makes it necessary to apply new and explicit methods [16]. The argument to skip the design process models is that they are not useful and do not represent what happens, in practice. Or as Gedenryd [11] sums up in a number of studies:

> On the one hand, (Design Process Models) do not work as prescriptions. People do not use them, because they do not work for their advertised purpose; those who actually tried them failed to reach the stated results. On the other hand, they are also inadequate as descriptions. If you study how practitioners really work, you will find what they really do to be something quite different (p. 66).

Parallel with the discussion on design process models in the design community, several studies have been made with designers in practice and design as something separate from science. An e.g of this is Bryan Lawson's book: *How Designers Think* [20].

In this book, Lawson argues that designers are very different from scientists in that scientists set out to study the problem, whereas designers learn about the problem as a result of trying out the solution. This means that designers co-develop the understanding of the problem along with the creation of the solution. According to Lawson, this also means that designers are more inclined to generate a fairly quick and satisfactory solution, rather that prolonging the analysis of the problem.

This is also related to the kind of reasoning deployed in design. Roozenburg and Eekels [29] argue that design reasoning is abductive. Instead of building an hypothesis, which can be tested (deductive reasoning) or gathering a large set of inquiry, on which a rule or argument can be based (inductive), designers start off with a set of seemingly unrelated facts, sensing that they are somehow connected. Both the solution and the hypothesis emerge as an end result of connecting these facts.

Another significant study of designers in practice was made by Donald Schön. In his book 'The reflective Practitioner', Schön argues that design is a 'reflective conversation with the situation' [31]:

> (…) I shall consider designing as a conversation with the materials of a situation. A designer makes things. (…)
> He works in particular situations, uses particular materials, and employs a distinctive medium and language. Typically, his making process is complex.
> There are more variables—kinds of possible moves, norms, and interrelationships of these—that can be represented in a finite model.
> Because of this complexity, the designer's moves tend, happily or unhappily, to produce consequences other than those intended.
> When this happens, the designer may take account of the unintended changes, he has made in the situation by forming new appreciations and understandings and by making new moves. He shapes the situation in accordance with his initial appreciation of it, the situation "talks back," and he responds to the situation's back-talk. In a good process of design, this conversation with the situation is reflective. In answer to the situation's back-talk, the designer reflectsinaction on the construction of the problem, the strategies of action, or the model of the phenomena, which have been implicit in his moves ([31], p. 79).

2.1.4 The Design Perspective Versus the Early Phases of Innovation

As described in the introduction chapter, interdisciplinary teams working in the early phases of innovation are challenged by (1) Diversity, (2) Complexity and Ambiguity, and (3) Asymmetry or Stickiness of information. These challenges can also be found in design projects in general—if not all at once then at different times in the projects. This indicates that some of the approaches and perspectives from design may be relevant to use in projects in the early phases of innovation. This argument can also be supported by the characteristics of the designer, which can be summed up on the basis of the review above:

- Designers tackle wicked problems and approach all problems as if they were ill-defined.
- Designers are human-centered and have a myriad of tools and methods to approach the user (or the network of stakeholders).
- Designers co-develop the understanding of the problem along with the creation of the solution.
- Designers use abductive reasoning and strive for a solution.
- Designers engage in a reflective conversation with the situation.

All of these characteristics or attributes seem relevant, when it comes to approaching the challenges in the early phases of innovation.

In relation to the challenge of diversity in the interdisciplinary team, the design perspective is relevant because of its human-centeredness. The human-centeredness is a focus point, which is shared by all the team members and stakeholders, irrespective of their background or perspective. Furthermore, the aim in the interdisciplinary team working in the early phases of innovation is to find the right problem, need, or opportunity, and to accomplish this a human-centered approach is very relevant.

In relation to the challenge of complexity, the design perspective is relevant, because of its ability to tackle wicked problems and the co-development of problem understanding and solution. First of all because the problems in the early phases of innovation are wicked, and often ill-defined. And second, because one of the most plausible ways to handle the complexity in the early phases is by co-developing the understanding of the problem along with the creation of the solution—in a reflective conversation with the situation.

In relation to the challenge of ambiguity, the design perspective is also relevant, because of its abductive reasoning and reflective conversation with the situation. In the early phases of innovation, there is no causality, and therefore abductive reasoning becomes a relevant approach along with probing different possibilities.

And finally, in relation to the asymmetry or stickiness of information, the design perspective is relevant, because of its human-centeredness and the myriad of methods with which to approach the user (or the network of stakeholders).

Another link between the early phases of innovation and the design perspective can be found in practice, where design is already playing a significant role in the early phases of innovation [10, 44], and where several researchers from the field of business strategy and management praise the use of design approaches in early phase innovation projects [21, 22].

In Chap. 3, there will be a more detailed explanation of how the different insights and approaches from the design perspective are adopted and used in relation to the workshops and the research, in general. In the following section, however, the focus will be on the theoretical framework.

2.2 Theoretical Framework

As explained in the beginning of this chapter, the theoretical framework is focused on meaning making. In the literature, meaning making is examined in a number of areas for instance in leadership, teaching, organizational learning, religion/spirituality, etc. Meaning making as a phenomenon derives from the hermeneutics; however, in this book, the intention is to view meaning making in relation to the creation, design, and development of new products, processes, and services, and more specifically, in relation to individuals, teams, and physical artifacts involved in this process of creating, designing, and developing.

In the first section, meaning will be reviewed in relation to how the individuals are creating meaning and how this meaning is communicated. This is relevant to the research, because each member of the design team as well as each stakeholder will have their own way of making meaning in relation to the early phase project, which they are working on.

Second, meaning will be reviewed in relation to how teams are creating shared frames. This is relevant in relation to the research, in terms of the need to create a shared project frame or other aspects of 'sharedness' within the team which previous research has shown is important.

And finally, meaning will be unfolded in relation to physical artifacts, in terms of how they impact communication of meaning as well as the creation of shared frames.

2.2.1 Meaning in Relation to Individuals

> One cannot ignore that designers, engineers, business people, politicians, cultural critics, and users all live in different worlds, act according to different conceptions they bring to what they encounter, and create different meanings for what seems from any one perspective to be the same thing ([17], p. 49).

In order to understand the process of meaning making on the individual level, it is important to make a distinction between meaning and sense. Whereas sense is immediate, direct, and almost unconscious gathering of insights from the surroundings, meaning involves conscious reflection and interpretation. Sense happens throughout all of our senses, whereas meaning involves an intellectual molding [17]. The difference between sense and meaning can be explained through

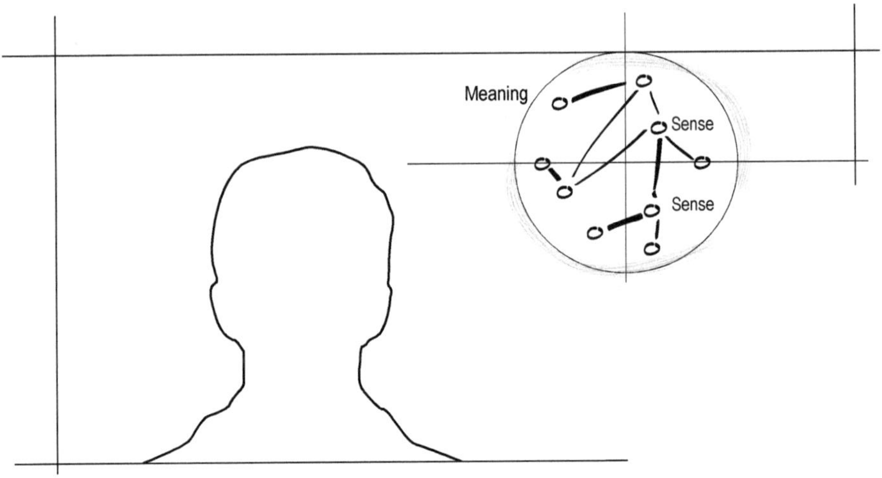

Fig. 2.5 Meaning as a network of senses

the metaphor of a puzzle: sense is the interdependent pieces, while meaning emerges when the different pieces are added together—into something meaningful.

> Sense is the feeling of being in contact with the world without reflection, interpretation or explanation. (…). Sense is the background against which one notices what is unusual, unexpected or different. Sense is the tacit, taken for granted and largely unconscious monitoring of what is ([17], p. 50).

The question of meaning is often brought up when something unexpected is sensed, or when an alternative way to combine senses is introduced, i.e., if you come home and something in your house has changed from its usual position, you will notice it. Likewise, if you see a person staggering toward you, you might think he is drunk—however, after talking to him you may realize that an aggressive-type sclerosis is the reason for the awkward type of walking, and this will probably change your perception and attitude toward him.

> Meaning is a structured space, a network of expected senses, a set of possibilities that enable handling things, other people, even oneself. They guide action much as a map shows all the possible paths from where one stands ([17], p. 56).

This connection between sense and meaning is also illustrated in Fig. 2.5. Each of the black dots represents senses and the whole figure illustrates the meaning.

2.2.1.1 The Links Between Sense and Meaning

As it has been revealed above, meaning can be seen as an explanation of how a sense is embedded in the context of other senses, and the senses' role in this context. Krippendorff [17] argues that meaning manifests itself in different ways:

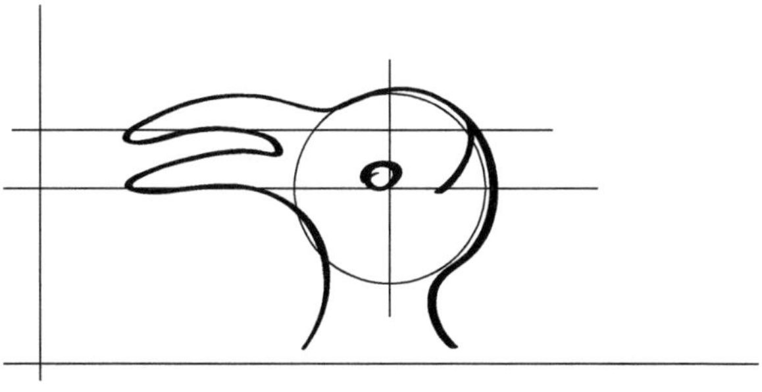

Fig. 2.6 The Duck-Rabbit [48]

in perception, in reading, in language, in conversation with others, and in re-representation. In relation to this book, the manifestations in perception and in conversation with others are especially interesting.

In perception, meaning arises in the awareness of the possibility of different ways of seeing [17]. So, when we notice that something can be perceived in more than one way, we become aware that we are making meaning of things in a certain way or from a certain perspective. A very simple and well-known example of this is the Duck-Rabbit created by Wittgenstein [48]. It can be seen in three ways: as a duck, as a rabbit, and as a line drawing (Fig. 2.6).

When looking at the drawing, it seems as if it is shifting from a duck to a rabbit, depending on how we perceive it. However, nothing in the drawing—as such—is changing. It is in our perceptions or meaning making that the changes occur.

In conversations with others: questions of meaning can also arise when we become aware that others seem to see things differently, when others use words or handle artifacts in ways we would not, or when others account for their world in terms different from our own. Experiencing such discrepancies challenges the obvious of our own perceptions, and accepting the possibility of versions other than our own calls for explanations of these apparent differences [17].

This implies our personal meaning making becomes challenged, when we realize that other people see the world differently, and it sharpens our consciousness in relation to how we construct meaning of our own.

In Fig. 2.7, perception and conversation with others are illustrated as links between sense and meaning, or as manifestations of two different meaning-making processes.

2.2.1.2 Second-Order-Understanding

Sense and meaning are both 100 % personal: they can never be completely shared with others, because they are based on personal experience.

However, in the early phases of innovation, it is important for the interdisciplinary team to obtain an understanding of the meaning, which stakeholders apply to the situation or activity in question as well as to understand how the other team

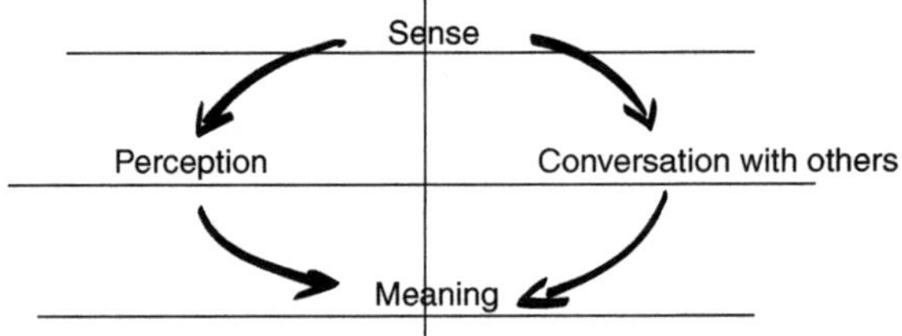

Fig. 2.7 The link between sense and meaning

members apply meaning. This kind of understanding is often referred to as second-order-understanding.

> Understanding someone else's understanding is an *understanding of understanding*, an understanding that recursively embeds another person's understanding in one's own, even if, and particularly when, these understandings disagree, contradict one another, or are thought by one to be wrong or appallingly unethical. This recursive understanding of understanding is a second-order-understanding. ([17], p. 66, original emphasis)

Narratives and stories are typical strategies to gain a second-order-understanding, because these already play an important role in people's understanding, explanations, and representations of life. On the basis of previous research, Cochran [9] has summarized the following links between the narrative and its influence and importance to human life.

> First, we live in story. In human experience there is always before and after, memory and anticipation, without which the present is unintelligible. (...)
> Second, we represent life in story. (...) In telling jokes, giving anecdotes, describing an event, writing a life story, or planning the future, we tell stories. (...)
> Third, we explain through story. (...) explanation is not concerned with one time or another, but with the change over time represented by the two contrasting states. (...). An explanation takes the form of a story, because it already has the story's form, with a beginning, middle and end.
> Last, we understand and comprehend through story. (...) To comprehend, we seek larger patterns and syntheses in which parts fall into place ([9], p. 73).

However, when looking at the situation in the early phase of innovation, for many stakeholders it is simply impossible to narrate or explain how they apply meaning to a situation or activity, either because they have done it so many times that they do not think of it anymore, or because they find it hard to define or articulate.

Likewise, in the interdisciplinary team it can be difficult to understand the meaning which the other team members apply, because each team member has a certain perspective, professional language, and set of experiences.

In many cases it is therefore relevant to enter into a dialog and re-examine how each individual experiences, senses, feels, and behaves in certain situations and investigate how these small pieces of information are provided with meaning in their 'worlds'.

> While stories can never capture all the meaning that informants bring into a narrative, especially their feelings and tacit understandings, conversation provides a window into the understanding that others have (…). The key to this understanding is unprejudicial listening, avoiding our own categories, and being careful rearticulating these stories in our own terms ([17], p. 55).

2.2.1.3 Metaphors

According to Lakoff and Johnson, metaphors provide another way for people with different values, experiences, and perspectives to gain second-order understandings. In their book 'Metaphors we live by' they state that:

> A metaphoric presentation skill is essential to create contact and communicate experiences, which are not shared ([18], p. 257).

This means that metaphors become a way to explain values and meanings, as well as a way to frame and restructure understandings. It is therefore interesting to look more closely at present research on metaphors. It is evident that metaphors function by explaining something about an object, activity, or relation in terms of something else. For instance in the metaphor: "there is chemistry between them", the world of chemistry tells us something about how to understand the relationship between the two people in question. On the more operational level this can also be explained like this:

> Metaphors operate across two logically independent domains, an absent but familiar domain of experience, the source domain, and a present domain in need of understanding or restructuring, the target domain ([17], p. 157).

In relation to the example 'there is chemistry between them', this means that the source domain is chemistry, and the target domain is the relationship between two people. It is further found that metaphors transfer a way or understanding or a pattern of understanding between the two domains:

> Metaphors carry (…) patterns of understandings embedded in the vocabulary of source, from the source domain along the structural resemblance into the target domain, which becomes reorganized regardless of what it was previously ([17], p. 157).

In the example used above about chemistry between two people, the patterns of understanding in chemistry are transferred to the domain of human relationships. However, metaphors are not just interesting in relation to communication and the creation of second-order-understandings, they also have a very significant influence on our cognition. Lakoff and Johnson argue that:

> (…) most of our ordinary conceptual system is metaphorical in nature ([18], p. 4).

This means that we use metaphors as a way of mapping our experiences into our cognition and thereby making meaning of them. Their research shows that we perceive, understand, and structure things in terms of other things. Even if we do not use the conceptual metaphors directly, it is revealed in the way we think and talk about things. For instance, the conceptual metaphor, Love is Magic, can be found in a number of everyday expressions like:

> She cast her *spell* over me. The *magic* is gone. I was *spellbound*. She had me *hypnotized*. He has me in a *trance*. I was *entranced* by him. I'm *charmed* by her. She is *bewitching* ([18] p.62, original emphasis).

The use of metaphors as a tool to structure our experiences becomes particularly important, when it comes to comprehending things, which cannot be comprehended totally like feelings, moral practices, and spiritual awareness [18]. But the argument that our conceptual system is metaphoric in nature also emphasizes the value of communicating through metaphors.

2.2.1.4 Summary With Respect to Individual Meaning Making

In the section above it has been reviewed how individuals are making meaning and how this meaning making can be communicated to others. It was found that sense is immediate, direct, and almost unconscious gathering of insights from the surroundings, whereas meaning involves conscious reflection and interpretation. It was further found that meaning arises both in perception and in conversation with others.

In relation to the communication of meaning, it was found that when meaning is communicated, it is only possible to create a second-order-understanding of the meaning shared. And that conversation and metaphors were identified as possible ways of creating this second-order-understanding, because both of these are able to capture feelings, moral practices, and tacit understanding, which are very hard to communicate. And finally, metaphors were especially highlighted because we structure our senses and make meaning of things via the use of metaphors or metaphoric features.

The theoretical framework on individual meaning making, which has been reviewed in the section above, is relevant in relation to this study, because:

1. It creates insights into how each team member in the design team is making meaning in relation to the project.
2. It creates insights into how each stakeholder or user is making meaning in relation to their everyday life and
3. It provides insights into how this meaning making can be communicated.

2.2.2 *Meaning in Relation to Teams*

One of the main findings in relation to meaning making is that both sense and meaning are 100 % personal. Therefore, it does not really make sense to talk about shared meaning within the interdisciplinary team. Instead, it is possible to work with 'sharedness' in terms of shared frames. Framing as a concept has received attention from the fields of sociology, urban planning, engineering, linguistics, cognitive science, management science, and organizational behavior. Still, there is no comprehensive or definite definition of it [39]. The presentation of team or project framing will be preceded by a brief review of framing in general, which often is defined in relation to the individual. However, even if frames can be understood in relation to individuals and organizations, this book will mainly use the understanding of frames in relation to teams or projects.

2.2.2.1 Framing

Framing as a phenomenon originates from perception psychology and refers to the organization of experiences [12]. All individuals build their own set of framings based on their everyday experiences. However, the framing does not only function as a means to organize information, but also as a filter for screening of all incoming information. This means that the already existing frames shape the 'rules' or guidelines in relation to the perception of upcoming situations [36]. Accordingly, to change an individual's understandings and thereby actions taken, one has to change the person's frames. Argyris [1] explains the creation and use of framings through his ladder of inference viewed in Fig. 2.8.

On the first step of the ladder the observable data or experiences are placed. This is then going through a selection process on the second step of the ladder. The selected data and experience are then structured into meaning and processed to assumptions. On the basis of this it now becomes possible to draw conclusions and at some point adopt beliefs about the world. These beliefs are the basis of actions, but also an active player in the data, which is selected next time. Going through the ladder of inference also means that a frame is created.

In her thesis, Valkenburg defines framing as a device for sense making, which settles the parameters of the problem [45]. This perspective on the frame is identifiable in the work of Weick [47], whose research has been focused on sense making in organizations.

One of the most comprehensive studies and discussions on frames and framing has been developed by Schön. He has described frames as:

> underlying structures of belief, perception, and appreciation ([32], p. 23).

And framing as an activity, in which the aim is to construct meaning. He further concluded that frames include implicit assumptions about what issues are relevant, what values and goals are important, and what criteria can be used to evaluate success.

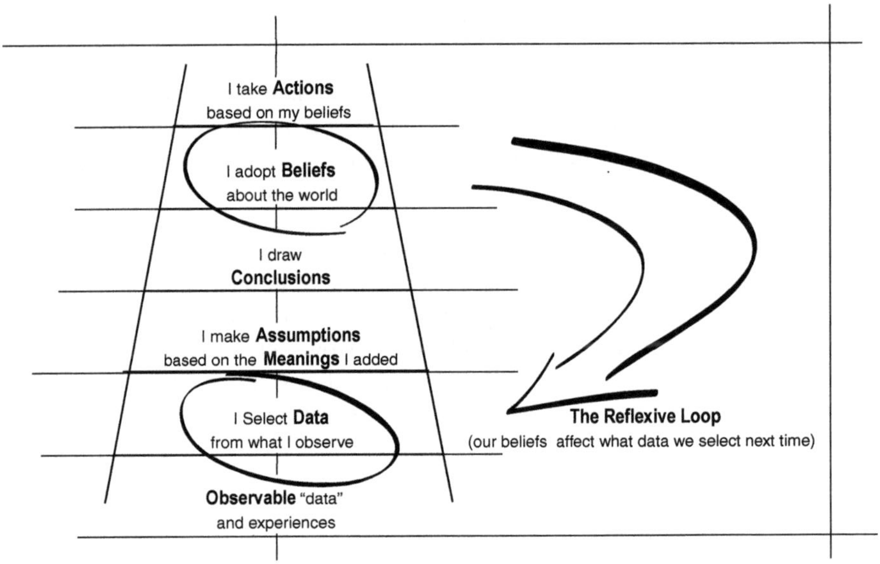

Fig. 2.8 Ladder of inference (Illustration from [34] p.243)

> Problem setting is the process in which we name things to which we will attend and frame the context in which we will attend to them ([31], p. 40).

In his book The Reflective Practitioner, Schön [31] describes the contrasting frames a design student and his mentor bring into play in a design review, and how their interactions with each other and the framing of the problem produce the final design. In the review, the names of the design strategies were highlighted in particular as well as the importance of shifting frames, when a problem becomes problematic to handle within one framing. Furthermore, Schön found generative metaphors as a means to frame situations—and thereby also frames as a way of 'seeing as' [31]—that is, seeing something through a lens of something else. He gives an example of this in terms of a framing used in an urban planning project, where a slum area either can be framed as a 'blighted area', which needs to be cured or as a 'natural community', which should be preserved. Both frames represent a significant view of the design situation and evoke different understandings of the problem.

2.2.2.2 Team Framing and Negotiating Shared Frames

Valkenberg and Dorst [46] attempted to apply the thinking in Schön's book: The Reflective Practitioner to team design situations. In practice, it involved a study of industrial design students working on projects in teams. When comparing the successful team with the unsuccessful team, it was found that the team's problem framing played a significant role. In the successful teams, it was possible to

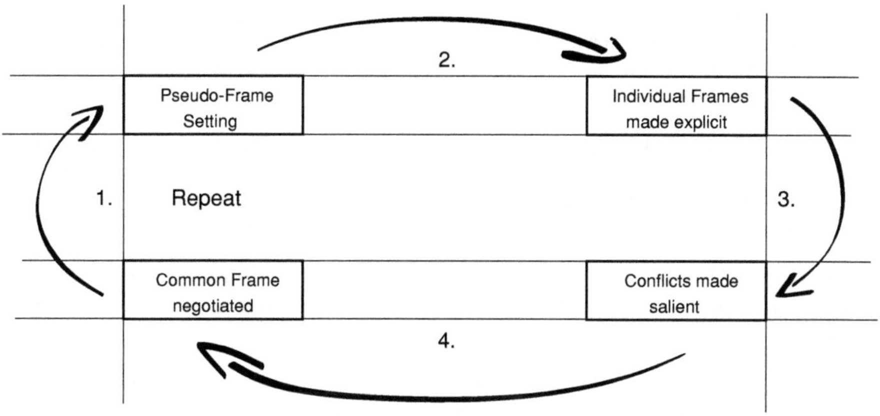

Fig. 2.9 The four phases of the framing cycle [15]. Own illustration

identify five different frames used sequentially, whereas in the unsuccessful team only one single frame was used.

In relation to teams, the project frames can be defined as the basis upon which the team pairs problem with solutions [31, 46] this means the selection of a desired end state or goal, which implicitly includes the problem, need, or opportunity, or the identification of the problem, need, or opportunity, which implicitly includes the desired end state or goal.

In 2007, Hay et al. studied the creation of framing and reframing in design teams working in the early phases of product development projects. They found that project framing and reframing in design teams happen in four different phases. The four phases illustrated in Fig. 2.9 are reviewed below.

> *Pseudo-frame setting*
> Pseudo-frame setting creates an initial understanding of the design situation, the goals, important features, boundaries and evaluation criteria, in this case upon the initial presentation of the project proposal to the class. The proposal sets boundaries around the problem and solution domains and, in some cases, implies strong solution directions. (…) The use of broad, abstract language leads members to believe that they are on the same page. However, the vague nature of these initial agreements can mask deep-level disagreements in goals, assumptions, values and understandings. Individual frames are still hidden ([15], p. 93–94).

> *Individual frames made explicit*
> The process of interacting with and collecting data from users breaks down designers' preconceptions by informing their point of view and challenging their assumptions. (…) In the process of making these decisions, members often discover their own implicit ideas regarding the project. The consequent sharing of expectations through team interaction makes each member's implicit frame more explicit, and thus tractable material for discussion and debate ([15], p. 94).

> *Conflicts among individual frames made salient*
> Several activities make individual frames explicit, and thus conflicts among them salient, e.g. building a group vocabulary through defining terms, writing mission statements and

> other textual artifacts, labeling user needs and other concepts, prioritising user needs, categorising ideas and dimensioning users. Once conflicts between individuals' frames are made salient, common frames can begin to be negotiated ([15], p. 94).
>
> *Common frame negotiated*
> Teams that made individual frame conflict salient used a combination of user data, discussion, and listening to negotiate a shared frame. Dedication of the team to the user-centered design gives the team a common anchor against which to tether constantly evolving individual frames, thus enabling the eventual arrival at a shared frame ([15], p. 95).

Often it was found that the four phases of framing were repeated in the projects. Still, it was found very useful to go through a full iteration as soon as possible in the project:

> The sooner the team was able to 'get on the same page', as many students described it, the sooner they were able to focus on addressing the needs of their users without differences in understanding and assumptions getting in the way ([15], p. 95).

2.2.2.3 Summary With Respect to Meaning Making in Teams

In the section above, meaning has been reviewed in relation to teams. It was found that it did not really make sense to talk about shared meaning within a team, since both sense and meaning are 100 % personal. Instead it was found possible to work with 'sharedness' and meaning making in terms of shared frames. The review showed that framing can be seen as a device for sense making, which settles the parameters of the problem, and that a frame can be described as underlying structures of belief, perception, and appreciation. It was furthermore found that framing can be seen as an activity, the aim of which is to construct meaning.

In relation to team or project framings it was found that frames include implicit assumptions about what issues are relevant, what values and goals are important, and what criteria can be used to evaluate success. And finally, it was found that project framing and reframing in design teams happen in four different phases: (1) Pseudo-frame setting (2) Individual frames made explicit (3) Conflicts among individual frames made salient and (4) Common frame negotiated.

The review on frames and shared framing is relevant in relation to the book, in terms of the need to create a shared project frame within the team working in the early phases of innovation as well as a means to create the 'sharedness' within the team, which previous research has shown is important.

2.2.3 Meaning in Relation to the Creation of Physical Artifacts

The last area, which is reviewed in relation to meaning, is physical artifacts. Physical artifacts—including models, mock-ups and prototypes—are typically used throughout the development process as a tool to visualize the problem, the solution or parts of these. In other words, physical artifacts are playing significant roles, when it comes to combining insights and meaning making in design and development teams.

Some researchers even describe product development as a modeling activity, where the progression between models with different purposes drives the development process, and where the models become an important tool to describe, visualize, and sculpture one's thoughts as well as designing or communicating with others [5].

Schön [31] has also studied modeling and prototyping as part of the design and development process. He argues that when a designer works on the model, he encounters a number of unexpected challenges in the model and responds to these immediately, by using his tacit knowledge-in-action, which is based upon previous experiences. This is done in a process of trial and error, or what Schön has named reflection-in-action.

Michael Schrage [33] has studied modeling and prototyping in teams. In his book "Serious Play" he argues against the common assumption that 'great teams make prototypes' and suggests instead that 'prototypes make great teams'.

> In this sense the values of prototypes reside less in the models themselves than in the interaction—the conversations, arguments, consultations, collaborations—they invite. Prototypes force individuals and institutions to confront the tyranny of trade-offs ([33], p. 20).

Schrage further argues that prototypes can promote the awareness and empathy between collaborators within cross-functional and cross-disciplinary teams and work as a shared medium of communication and collaboration in the innovation process. In his perspective, prototypes can be seen as a tool, which minimizes the competition and discussions within the team and instead creates a place for collaboration [33]. This is supported by Henderson, who argues that prototyping is more than communication and coordination, and that it plays an important role in terms of the 'social glue':

> The analysis reveals that visual representations, including prototypes, are not only devices for communal sharing of ideas but are also a ground for design conflict and company politics, exactly because they facilitate the social organization of workers, the work process and the concepts that workers manipulate to produce a collective product ([14], p. 10).

2.2.3.1 Definitions: Models, Prototypes, and Mock-ups

The term model can be seen as the overall umbrella in which other more specific definitions can be found, such as prototypes and mock-ups. According to Schrage [33]:

> A model can be anything from a mathematical equation scribbled on a napkin to a full-scale version of a Boeing 777 (p. 7).

Within the definition of a model there are the definitions of prototypes and mock-ups. These will be reviewed below.

2.2.3.2 Prototypes

Prototypes can—just as models—take many different forms. A prototype can for instance be a scale model of a house or product, a piece of software, a paper-based outline of one or more screens, a video-simulation of a work task, or a three-dimensional mock-up of a workstation [26]. According to Preece et al. [26]:

> A prototype is a limited representation of a design that allows users to interact with it and to explore its suitability (p. 241).

Furthermore, prototypes can be used in a variety of ways. They can support designers and stakeholders to choose between different design alternatives, they can be used to test technical aspects of an idea or concept, and they can help to clarify requirements, test usability, or check if a certain design direction is in line with other parts of the design [26].

Preece et al. have divided prototypes into two categories: low-fidelity prototypes and high-fidelity prototypes. Low-fidelity prototypes are often made of simple and cheap materials like paper and cardboard. They are often cheap, fast to produce, and modify. As a result, low-fidelity prototypes are different from the final design, and therefore they are not to be kept and integrated into the final product.

High-fidelity prototypes look more like the final design, and they are made of the same materials as the final design. High-fidelity prototypes are more time-consuming and hereby more expensive than low-fidelity prototypes.

In relation to collaborative design, Bødker and Buur [6] stress the importance of tangible prototypes as tools to try out future use situations, because one can interact with them and get hands-on experiences. However, there is a limit to the meaning they convey, or as Shaw expresses it:

> Prototypes make very definite statements about the precise nature of what is envisioned and allow these to be tested in the context of use, but do not by themselves convey the reasoning behind any particular feature or alternatives that may have been considered ([35], p. 70).

2.2.3.3 Mock-ups

Mock-ups belong to the low-fidelity category and have been described in relation to various design and developments contexts. According to Merriam Webster's dictionary a mock-up is:

> -a full-sized structural model built to scale chiefly for study, testing, or display, or
> -a working sample (as of a magazine) for reviewing format, layout, or content. [25]

Carroll [8] has studied the use of mock-ups in scenario-based design, and Binder [2] has studied how users with simple cardboard mock-ups as props can create improvised scenarios in their own environment.

2.2.3.4 Boundary Objects

Many models, prototypes, and mock-ups can also be described in terms of boundary objects. The concept of Boundary Objects is described by Carroll [38], defining objects that are shared and sharable in different problem solving contexts, that is objects which work to establish a shared context or which 'sit in the middle'.

> Boundary objects are objects which are both plastic enough to adapt to local needs and constraints of the several parties employing them, yet robust enough to maintain a common identity across sites. They are weakly structured in common use, and become strongly structured in common use. These objects may be abstract or concrete. They have different meanings in different social worlds, but their structure is common enough to more than one world to make them recognizable as means of translation ([38], p. 393).

Star and Griesemer [38] identified Boundary Objects in a museum context, where, in the beginning of the 1900s and onwards, scientists and amateur collectors used Boundary Objects to create a shared collection of Vertebrate material, which could be useful to both communities. Based on this study Star and Griesemer found that:

> The creation and management of boundary objects is a key process in developing and maintaining coherence across intersecting social worlds ([38], p. 393).

According to Henderson, the most important aspect of Boundary Objects is that they make it possible for different groups to see and understand different meanings in the objects. Boundary Objects shall thus be understood as objects that can give meaning to different participants, even though they have different professional practices and professional languages—different competencies.

> (…) Boundary Objects allow members of different groups to read different meanings particular to their needs from the same material. This is possible, because the material remains flexible in group use and more focused in individual site use [13].

According to Miller [24] it is not only important that the Boundary Objects are created, but that they are co-invented, developed in neutral territory, have a reasonable lifespan, and have real use and meaning to all the participants. Carlile [7] have studied the difference between good and bad Boundary Objects or, in other terms, the difference between 'Boundary Roadblocks' and Boundary Objects. On the basis of this he has identified the characteristics of a good Boundary Object as follows.

Good Boundary Objects:

1. Establish a shared syntax or language for individuals to represent their knowledge.
2. Provide a concrete means for individuals to specify and learn about their differences and dependencies.
3. Facilitate a process where individuals can jointly transform their knowledge.

However, as Subrahmanian et al. argue:

> Boundary Objects can inhabit several communities of practice and satisfy the information requirements of each of them. This does not mean that use of Boundary Objects requires participants to have shared understandings to establish coordination ([40], p. 186).

2.2.3.5 Summary of Meaning Making in Relation to Physical Artifacts

In the section about meaning in relation to physical artifacts it was found that models and prototypes are playing significant roles, when it comes to combining insights and meaning making in design and development teams.

It was further found that prototypes can promote the awareness and empathy between collaborators, work as a shared medium of communication, minimize the competition and discussions within the team and function as the 'social glue'. The definitions of models, prototypes, and mock-ups were reviewed as well as some of the more individual characteristics.

It was further found that models, prototypes, and mock-ups can play the roles of Boundary Objects, which makes it possible for different groups to see and understand different meanings in the same objects. Boundary Objects as a phenomenon was also further elaborated and defined, in terms of objects that are shared and sharable in different problem solving contexts. And further research reviewed showed it to be important that Boundary Objects are co-invented, developed in neutral territory, have a reasonable lifespan, and have real use and meaning to all the participants.

2.2.4 Connecting the Theoretical Framework to the Empirical Setup

In the review of the theoretical framework, which has been presented in this chapter, it was necessary to divide the meaning making into three directions: individual, team, and artifacts. However, in the workshops, on which this book is based empirically, the different types of meaning and the communication of meaning will happen simultaneously. That is: both individual meaning making, communication of meaning, creation, and negotiation of shared frames will happen at the same time.

On top of this, the workshops will also include the creation of physical artifacts. As explained in the preface, this book builds on the initial assumption that the creation of physical artifacts can help team-members, users, and stakeholders to overcome the boundary of not being able to define, express, and communicate how they frame a given project or make meaning in relation to their everyday life. And that this clarity will help the creation of a shared frame. The literature in the theoretical framework seems to underline this assumption, and it indicates that the

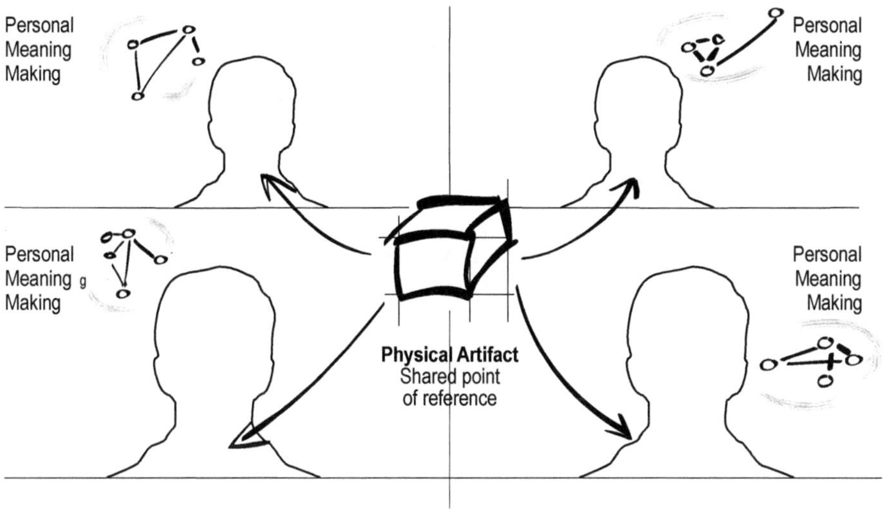

Fig. 2.10 Initial understanding of the workshop setup

physical artifacts created by the team members, users, and stakeholders could perhaps function as boundary objects, or at least function as a shared reference point for the team. This is also illustrated in Fig. 2.10.

References

1. Argyris C (1990) Bryt forsvarsrutinene—Hvordan lette organisasjonslæring. University Press, Cambridge (Original title: Overcoming Organizational Defences: Facilitating Organizational Learning)
2. Binder T (1999) Setting the stage for improvised video scenarios. In: CHI'99 Extended Abstracts (Pittsburgh, PA, May), ACM Press, New York, pp 230–231
3. Bucciarelli LL (1994) Designing engineers. MIT Press, Cambridge
4. Buchanan R (1992) Wicked problems in design thinking. Des Issues 8(2):5–21
5. Buur J, Andreasen MM (1989) Design models in mechatronic product development. Des Stud 10(3):155–162
6. Bødker S, Buur J (2002) The design collaboratorium—a Place for usability design. ACM Transactions on Computer-Human Interaction (TOCHI) 9(2):152–169
7. Carlile PR (2002) A pragmatic view of knowledge and boundaries: Boundary objects in new product development. Organ Sci 13(4):442–455
8. Carroll JM (2000) Making use—scenario-based design of human-computer interactions. The MIT Press, Cambridge
9. Cochran LR (1990) Narrative as a paradigm for career research in Young R A and Borgen W A methodological approaches to the study of career. Praeger Publishers, New York
10. Friis SK (2006) Conscious design practice as a strategic tool. Dissertation, Danish School of Education, Copenhagen

11. Gedenryd H (1998) How designers work—making sense of authentic cogni-tive activities. Dissertation, Lund University, Lund. http://www.lucs.lu.se/Henrik.Gedenryd/HowDesignersWork/
12. Goffman E (1974) Frame analysis. An essay on the organization of experience. Harper and Row, New York
13. Henderson K (1991) Flexible sketches and inflexible data bases: Visual communication, conscription devices, and boundary objects in design engineering. Science Technology, & Human Values 16(4):448–473
14. Henderson K (1999) On line and on paper: Visual representation, visual culture, and computer graphics in design engineering. MIT Press, Cambridge
15. Hey JHG, Joyce CK, Beckman SL (2007) Framing innovation: Negotiating shared frames during early design phases. J Des Res 6(1–2):79–99
16. Jones JC (1970) Design methods: seeds of human futures. Wiley, London
17. Krippendorff K (2006) The semantic turn—a new foundation for design. CRC Press/Taylor and Francis Group, New York
18. Lakoff G, Johnson M (1980) Metaphors we live. University of Chicago Press, Chicago
19. Laurel B (2003) Design research methods and perspectives. The MIT Press, London
20. Lawson B (1980) How designers think—the design process demystified. Architectural Press, London
21. Liedtka J, Mintzberg H (2006) Time for design. Des Manag Rev Spring 2006:9–18
22. Martin R (2004) The design of business. Rotman Manag (the alumni magazine) winter 2004:7–10
23. Merholz P, Wilkens T, Schauer B, Verba D (2008) Subject to change: Great products and services for an uncertain world. O'Reilly Media, Sebastopol
24. Miller R (2005) Creating boundary objects to aid knowledge transfer. Knowl Manag Rev 8(2):12–15
25. Merriam-Webster (2012). www.merriam-webster.com/dictionary
26. Preece J, Rogers Y, Sharp H (2002) Interaction design: Beyond human–computer interaction. Wiley, New York
27. Rittel HWJ (1972a) On the Planning Crisis: System analysis of the first and second generations. Bedrifts Økonomen (Norway) No. 8, pp 390–396
28. Rittel HWJ (1972) Second generation design methods. DMG 5th anniversary report, DMG occasional paper 1:5–10
29. Roozenburg N F M & Eekels J (1991) *Product design: fundamentals and methods* John Wiley & Son Ltd, London
30. Sanders L (2006) Design research in 2006. Des Res Quart 1(1):1–8
31. Schön DA (1983) The reflective practitioner. Basic Books, New York
32. Schön DA, Rein M (1994) Frame reflection: Toward the resolution of intractable policy controversies. Basic Books, New York
33. Schrage M (2000) Serious play—how the world's best companies simulate innovation. Harvard Business School Press, Boston
34. Senge P, Ross R, Smith B, Roberts C, Kleiner A (2001) The fifth discipline—fieldbook. Nicholas Brealey Publishing, London
35. Shaw B (2007) More than the sum of the parts: Shared representations in collaborative design interaction. Dissertation, Royal College of Art, London
36. Shazer S (1988) Clues, Investigating solution in Brief Therapy. Norton & Company, New York
37. Simon HA (1969) The sciences of the artificial. MIT Press, Cambridge
38. Star SL, Griesemer JR (1989) Institutional ecology, translations and boundary objects: Amateurs and professionals in berkeley's museum of vertebrate zoology, 1907–1939. Soc Stud Sci 19:387–420
39. Stumpf SC, McDonnell JT (2002) Talking about team framing: Using argumentation to analyse and support experiential learning in early design episodes. Des Stud 23(1):5–23

40. Subrahmanian E, Monarch I, Konda S, Granger H, Milliken R, Westerberg A, The N-DIM group (2003) Boundary objects and prototypes at the interfaces of engineering design. Computer Supported Cooperative Work 12:185–203
41. Thomas JC, Carroll JM (1979) The psychological study of design. Des Stud 1(1):5–11
42. Tjalve E (1979) A short course in industrial design. Newnes-Butterworths, London
43. Ulrich KT, Eppinger SD (2005) Product design and development. McGraw-Hill Book Co., New York
44. Van Patter GK, interviewed by Peter J Bogaards (2005) The Infodesign Interview. p 36–41 http://www.informationdesign.org/special/vanpatter_interview.php
45. Valkenburg R (2000) The reflective practice in product design teams. Dissertation, Technical University of Delft, Delft
46. Valkenberg R, Dorst K (1998) The reflective practice of design teams. Des Stud 19(3):249–272
47. Weick KE (2001) Making sense of the organization. Blackwell Publishing Ltd, Malden
48. Wittgenstein L (1953) Philosophical Investigations. Blackwell Publishing Ltd, Oxford

Chapter 3
Workshop Cases and Method

This chapter holds review of the workshop method and the research material. First, facilitated workshop will be introduced in terms of background and development as well as of its practical application. After this, there will be a review of the six workshop cases, their structure, content, and their context. Chosen highlights and models will exemplify the results from the workshops.

3.1 Background for Choosing Method and Process

The initial idea in this research project was to introduce the creation of physical artifacts in interdisciplinary design teams working in the early phases of innovation. The inspiration for this derives from the later parts of both design and innovation processes, where models and prototypes are often used to support communication, to test possibilities, and compare ideas etc.

The creation of physical artifacts was seen as a way to create a shared reference point among the different perspectives and thereby create a point of departure for the shared framing. A second reason for introducing the creation of physical artifacts in early phase projects was to help the team members, users and stakeholders to overcome the boundary of not being able to express, what they find meaningful in relation to the project or how they make meaning of their everyday activities.

However, there were a number of demands, which had to be fulfilled in order to introduce the creation of physical artifacts in the early phases of innovation:

- The requirement for creating the physical artifacts has to be kept to a minimum in order to involve as many relevant users and stakeholders as possible (since it is not given that everyone in the interdisciplinary team or among users/stakeholders has modeling experience).
- The creation of the physical artifacts has to handle a great diversity in terms of different backgrounds and assumptions as well as kinds and levels of knowledge. And it has to contain both the complexity and ambiguity, which may appear in relation to the project.

L. Møller and C. Tollestrup, *Creating Shared Understanding in Product Development Teams*, DOI: 10.1007/978-1-4471-4180-8_3,

- Finally, the creation of the physical artifacts has to involve a shared experience or shared output—in order to build on the insights from previous research, in which 'sharing' is identified as an important tool to handle diversity, complexity, and ambiguity in teams.

3.2 Choice of Method and Process

The first process, which was tested as a possible way to create physical artifacts in interdisciplinary teams working in the early phases of innovation, was a consultancy process for business and organizations called LEGO® Serious Play TM (henceforth: LSP).

LSP seemed to fulfill the demands described above, and it was directly accessible in this study. However, LSP was not directly adopted. Since it was mainly developed for a business context, it was found necessary to develop a new set of LSP variations, which could be used in projects positioned in the early phases of innovation. This was done by adapting and also further developing some of the main features of the original LSP, based on the design perspective reviewed in Chap. 2. In the following sections, the original LSP will be reviewed in terms of development, background, as well as key features. After this, a review follows of how the design perspective was used to further develop LSP and to make it applicable to projects in the early phases of innovation.

3.3 LEGO® Serious Play: Development and Background

LSP is often described as an accidental spin off from LEGO® Group. David Gauntlett [3] explains the development of it as follows: Lego Serious Play developed out of a problem within the Lego company itself:

> (…) In 1996-97, Kjeld Kirk Kristiansen [owner of Lego and grandchild of its founder] was feeling disappointed that his staff meetings did not seem to be able to generate imaginative strategies for the future of the company. He knew that his employees were talented people, and so felt that some kind of tool was needed to unlock their imagination and creativity. During this time, he had discussions with Bart Victor and Johan Roos, both professors and consultants from the Swiss Business School Imd, who had seen this kind of situation elsewhere. Together they realized that a solution to Lego's problem might be found in the Lego product itself: just as Lego had been telling children to 'build their dreams' for decades, so perhaps adults could be asked to 'build' their vision for future strategy. (p. 129)

According to its developers, LSP is based upon four theoretical directions, which are referred to as 'The Science of Lego Serious Play'. The directions are: (1) Constructivism (2) Constructionism (3) Play and (4) Imagination [6]. In Table 3.1 there is a brief review of the four directions.

Table 3.1 The theoretical background or 'Science' of Lego Serious Play

Constructivism and Constructionism:
The notion of constructivism is developed by Jean Piaget and concerned with how learning takes place—especially for children. Piaget discovered that information and knowledge is not just absorbed, but actively built into knowledge structures. This means that incoming experience is combined with previous experiences into a set of active theories [4]. Seymour Papert extended the constructivism theory into what he called Constructionism. Apart from the building of knowledge structures, Papert argues that learning happens especially when engaging in creation of physical objects—for example, with clay or Lego bricks [7]. Papert's theories are often referred to as 'thinking with your hands' or 'learning by making'. Furthermore, Papert noticed that when people are making something with their hands, they are in a more deeply engaged state, compared to situations where they are solving abstract questions in their mind. This deep engagement, which Papert noticed, is what Mihaly Csikszentmihalyi later defined as 'flow'.
Play and Imagination:
Play is often associated with children; however, a growing literature is encouraging adults to engage in play as well. It is argued that surprising insights and innovative ideas are more likely to occur in playful environments and through playful behavior [8, 9]. In LSP play is defined as: "(...) a limited, structured and voluntary activity that involves imagination - that is an activity limited in time and space, structured by rules, conventions and agreements among the players, (...) and drawing on elements of fantasy and creative imagination. [6] p. 4)"
Imagination is seen as a central part of playing. Imagination can be divided into three categories: descriptive imagination, creative imagination, and challenging imagination. Below, these three categories are defined:
"Descriptive imagination not only reveals what is happening in the often confusing world out there, but it enables us to make sense of it and to see new possibilities and opportunities (...) Creative imagination allows us to see what isn't there. It evokes truly new possibilities from the combination, recombination or transformation of things and concepts.(...) Challenging imagination, often using deconstruction or sarcasm, overturns all the rules and wipes the slate clean. [6] p. 14-17)"
Especially, 'creative imagination' is seen as a central part of playing, in general. However, in relation to LSP the 'descriptive' and 'challenging imaginations' have significant roles, too. Besides the role of imagination, it is further argued that when adults are playing, it also involves (1) social bonding, (2) emotional expression, (3) cognitive development, and (4) constructive competition.

Today, LSP has been developed to cover a range of applications and is used in various situations and industries. Some of the applications are reviewed below:

- *'Real Time Strategy for Enterprise'* a sequence of activities in which participants build metaphorical models representing their organization and then combine these into a shared identity of the enterprise; then build 'agents' (any external entity which the organization may have to connect or deal with) and place these in a landscape in relation to their main model; then build different kinds of connections; then consider future scenarios; and ultimately arrive at 'Simple Guiding Principles', which emerge from the activity and help to make future decisions.
- *'Real Time Strategy for the Team'* a version more oriented toward team- building, in which participants begin by constructing models representing what they bring to the organization; then create a part of an identity perceived by colleagues which they have not included themselves; then they are asked to review what has been built so far and to build a model representing 'the feel of the team'; then to build connections showing how the parts of the team relate; then to reflect on the past ways of dealing with events to ultimately arrive at 'Simple Guiding Principles' for the team.

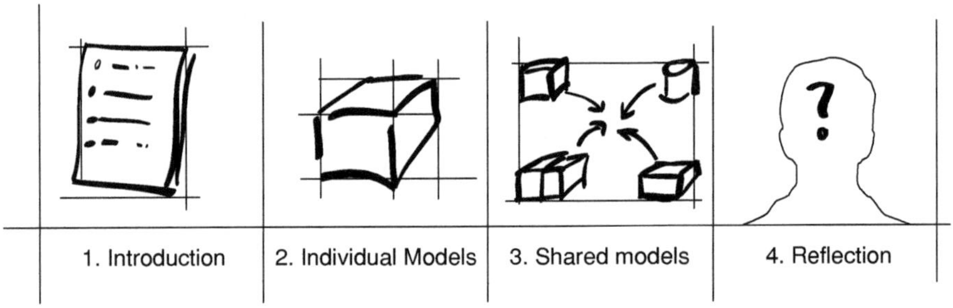

Fig. 3.1 The four parts of the LSP workshop

- *'Real Time Identity for You'* a simpler process in which individuals build a metaphoric model of their identity at work, then change it to show how they think they are perceived and then again to represent an aspirational version, 'what you could be at your best', and to reflect upon the differences. ([3] p. 135)

For more information about LSP's history and development please see the Lego Serious Play homepage (seriousplay.com), Imagination Lab (imagilab.org/research), or the book:'Creative explorations—new approaches to identities and audiences' [3]. However, be aware that some of the suggested material are marketing material and not research-based communication.

3.3.1 LEGO Serious Play is a Facilitated Workshop

In practice, LSP is a facilitated workshop, where participants are asked different questions in relation to an ongoing project, task, or strategy. The participants answer these questions by building symbolic and metaphorical models of their insights in LEGO® bricks and present these to each other. An essential part of the LSP is the nonjudgmental, freethinking, and somehow playful interaction among the participants [3]. A LSP workshop is divided into four parts. The first part is the introduction or 'skills building', where participants become familiar with the Lego pieces and the democratic process of building and presenting, as well as some of the key features like the hand-mind connection and the use of symbolic/metaphoric models. The next part is modified or tailored to fit each project, situation and context, in which it is used. This part often holds a subsession where participants first build a number of individual models and present these to each other; this is followed by a subsession where they build these models together into one shared model. And finally, there is a reflection and summary of the workshop in relation to the participant experience and future work. This is also illustrated in Fig. 3.1.

LSP can be seen to have similarities with a shared framing process, as described by Hay et al. [5] in the theoretical framework. First, the objective of the workshop is set, based on the present understanding of the project, task or strategy (pseudo frame setting), and the facilitation of the workshop is adjusted to fit this. The facilitator and the project manager often do this a few weeks before the workshop.

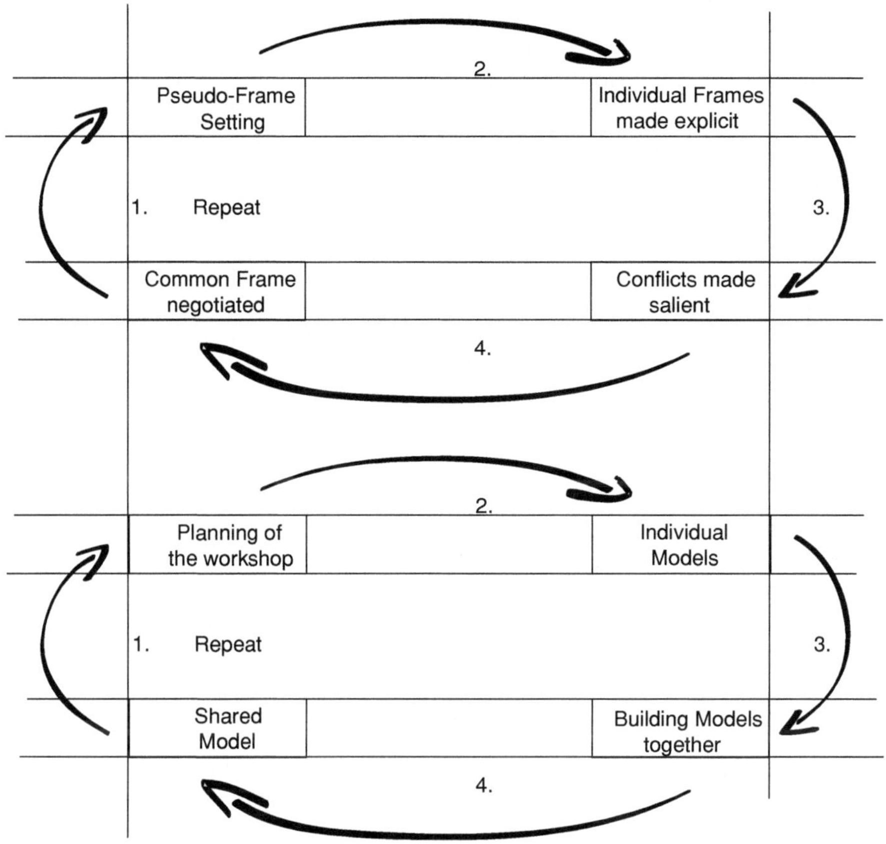

Fig. 3.2 The LSP workshop (*lower illustration*) as a simulation of a shared framing process (*higher illustration*)

Then, in the workshop (after the introduction/skills building) participants build an individual model, which represents their personal view on a question in relation to the project, task, or strategy. When presenting this individual model, they are making the insights and perspective from their personal model explicit to the other participants (individual meaning making made explicit).

In the next step of the workshop, participants are asked to build the Lego models together. In this process they are free to select one, a few, or all the models—and even to build new models—to be built together, as long as everybody in the group agrees that it is a shared model. In so doing, the conflicts among the different individual models become salient (conflicts made salient) and a shared model is created (common frame negotiated). This is also illustrated in Fig. 3.2.

Another important key feature in LSP is that each participant has equal opportunity to express his/her feelings or ideas and become part of the outcome, which emerges during the workshop. This means that the facilitator strives to give each participant the necessary time to explain the individual models, as well as the opportunity to become part of the 'building together' process. This is, for instance, done by controlling the line of speakers, when the individual models are presented and by asking clarifying questions about the models.

In the 'building together' process, this may be done by making sure that only one discussion is happening at a time and by reminding the group that everyone has to agree on the decisions taken in relation to the shared model. Another important feature of LSP is that the facilitator (consultant) is not providing 'the solution' to the project, task or strategy, or 'the answer' to the questions. Instead, the facilitator is concerned with the interactions between the participants, the coordination of the assignments, and the time.

3.4 The New Version

When this research project was initiated, there was no specific application of LSP, which could be used in the early phases of innovation. Therefore, the main focus in the first part of this research project was to design such an application. After some preliminary tests, it was found necessary not only to design one, but two different LSP applications, which could be applied in two different stages of the early phase projects.

1. The first application was only intended for the interdisciplinary team to be used as help to create a shared (and perhaps pseudo) framing of the project.
2. The second application included the interdisciplinary team as well as stakeholders and users, as a means to help the interdisciplinary team understand users' and stakeholders' perspectives on needs, problems, and opportunities in relation to the project.

The first application was called the team workshop, and the second one the stakeholder workshop. The reason behind the creation of two applications was mainly to keep the time frame of a single workshop down to a maximum of 8 h (with breaks)—and to be able to suit different projects in terms of their development.

3.4.1 The Design Perspective Influence

The creation of the two new workshops was accomplished by adapting and further developing some of the original features of LSP combined with insights and understandings derived from the design perspective (which was presented in Chap. 2).

The first perspective from design, which influenced the team workshop, was the designerly understanding of problems as wicked and ill-defined. In other words, it was assumed that the initial project descriptions in early phase projects were ill-defined and that the problems to be approached were wicked. In the team workshop, this can, for instance, be identified in the workshop questions. The workshop questions were often open-ended and asked in order to ascertain, whether participants had the same understanding of the project focus. An example of a question from one of the team workshops was: *What is the Good Elderly Life?* Another way to identify it was that even though many of the project managers asked for it, their workshop was never initiated by a presentation of the project, because this could indicate that there was one 'right' problem understanding or one 'right' solution.

Another feature from design influencing both applications, but especially the stakeholder workshop, was human-centeredness. This can be identified in different ways in the workshop. First of all, a broad set of stakeholders and users participated in these workshops. Secondly, the questions were often related to the users' and stakeholders' understandings and meaning making in relation to their everyday life. An example of a question from one of the stakeholder workshops was: *What are the challenges in a professional guitarist's life?*

A third perspective from design, which influenced the stakeholder workshop, was the co-development of the understanding of the problem parallel with the development of the solutions. In the stakeholder workshop, this is, for instance, identifiable in the combination of the questions, which included a question (or set of questions) with focus on the present problems and challenges, and a question (or set of questions) with focus on ideal solutions or future opportunities.

Besides the influence from the design perspective, both the team—and the stakeholder workshop include all the key features of LSP such as:

- The four parts (skills building, individual modeling, building together, and reflection).
- The democratic understanding of the process with equal opportunity to express insights, and
- The facilitator's focus on the interactions, assignment, coordination, etc., in relation to the workshop—and no interference with the content of the workshop.

The two new applications can also be summarized as follows:

1. *The team workshop*—a sequence of activities, where the participants in the interdisciplinary team reveal their different perspectives, perceptions, and assumptions about the project by building individual models as an answer to an open-ended question; then, the participants combine these into a shared model of the project. And together they figure out what is necessary to approach first, in the shared model. Ultimately, the workshop concludes with a reflection on its impact in relation to the project and the future development of this.

2. *The stakeholder workshop*—a sequence of activities, where the interdisciplinary team together with key stakeholders and users construct different individual models. First, the individual participants are asked to build a model that tells something about the main problems or challenges the users have (or are

assumed to have) in their everyday activities or life, in general. Then, the participants build an individual model, which represents the ideal solution or future opportunities. Finally, all the different models are combined into a set of *'guiding principles'* or *'design principles'* for the project—with respect to the user. The workshop is completed by a discussion of the outcome in relation to the project and in relation to the users' and stakeholders' experiences.

3.5 The Workshop Case Portfolio

In this section, there will be an overview of the workshop portfolio. This will be done by briefly presenting the workshops and identifying the general characteristics of the workshops and the projects, in which they are positioned. As mentioned earlier, the vehicle for this research is a set of video documented workshops with interdisciplinary teams, who are working on projects in the early phases of innovation. There are six workshops all in all. The first four are single workshops. This means that they are part of a larger project, but that the development of the project before and after the workshop has not been documented or cannot be shared due to confidentiality. The four single workshops were conducted at TC Electronic, Red Cross, Daimler AG, and Region Northern Jutland. Besides the single workshops, a longitudinal study of the user-driven-innovation project: 'The Good Elderly Life' has also been made, which included two sequential workshops and a more in-depth analysis of the project development. This means that the research is based on two different kinds of research material. On the one hand, there are four single workshops, which have been planned, executed, and documented as separate units of data. And on the other hand, there is a longitudinal study with two different workshops, multiple meetings, and access to information about the project development. Furthermore, the workshop portfolio holds both workshops, where the team workshop and the stakeholder workshop were used.

The choice between the two workshops was taken on the basis of the project and its development before the workshop. In Table 3.2, a more detailed overview of the workshop portfolio is presented. Here, all workshops are shortly described in terms of (1) organization or project in which the workshop takes place; (2) the project objective; (3) the workshop type; and finally (4) the material and type of data, which is available from the workshop.

When looking at the workshop portfolio, it is clear that the workshops are set in very different projects, contexts, and organizations. This is done to make the insights and conclusions deriving from the research as generalizable as possible.

Still, the conditions around the workshops are the same. They are all set in real-time projects, the workshops are actual parts of the projects and the participants in the workshops are real front-line workers. In other words: there is no time delay, no simulations, and the participants are professionals. These conditions are very important in relation to the reliability of the study in that it represents the actual situations in interdisciplinary teams working in the early phases of innovation.

Table 3.2 The workshops—in terms of context, workshop type and documentation

Organization/Project	Project objective	Workshop type	Material/ Type of data
Workshop 1 TC electronic	Development of digital sound solutions for guitarists.	Stakeholder workshop	Workshop/ Pictures and descriptions
Workshop 2: Red cross	Development of a Base Camp for international disaster management in situations like tsunamis or earthquakes.	Stakeholder workshop	Workshop/ Pictures and video
Workshop 3: Daimler AG	Incorporation of industrial psychology and team support into development teams	Team workshop	Workshop/ Pictures and video
Workshop 4: Region Northern Jutland *Medical Treatment House*	Development of guidelines for creating a set of medical treatment houses in northern Denmark.	Stakeholder workshop	Workshop/ Pictures and video
Workshop 5: Copenhagen Living Lab *The Good Elderly Life*	Development of new products and services, which can improve the life quality of elderly people living in nursing homes.	Team workshop	Workshop and longitudinal study of the project/Pictures, meeting minutes, informal interviews and video
Workshop 6: Copenhagen Living Lab *The Good Elderly Life*	Development of new products and services, which can improve the life quality of elderly people living in nursing homes	Stakeholder workshop	Workshop and longitudinal study of the project/Pictures, meeting minutes, informal interviews and video

3.5.1 The Workshop Cases Unfolded With Highlights

In this section, more life will be added to the six workshops, in the form of detailed descriptions of each of the workshops including pictures and quotes. The aim is to document the workshops in factual terms, while conveying as much of the workshop experience as possible.

As explained earlier all the workshops are simulating a shared framing process; initially, the workshops are planned and adjusted to the different projects based on the present understanding of these (pseudo-frame setting); then in the workshops, each participant builds an individual model, which is made explicit to the other participants (individual meaning making made explicit). And finally, participants are asked to build the Lego models together into one shared model, which means that conflicts between the models become salient (conflicts between the different frames made salient) and a shared model is created (common frame negotiated).

This structure of a framing process will also be the structure of the workshop presentation. First, the setting of the workshop will be presented including the context of the workshop, the workshop type, the assignment of the workshop, and the workshop participants. Then, the individual models will be presented. However, instead of presenting all the models (which in some workshops amount to about 20), only the models, which influence the shared model and which play a significant role in the workshop discussion, will be presented. When presenting the individual models, the conflicts among different models will also be presented. And finally, the shared model will be presented. On the following pages, the six workshops will be presented. The workshop descriptions follow the same standard. However, the presentation of the last two workshops (from the longitudinal study) will be more elaborate due to the extended data set.

3.5.2 Workshop Case 1: The Guitar Pedal, TC Electronic

This workshop was conducted as a Stakeholder workshop.

3.5.2.1 Context of the Workshop

Workshop 1 was conducted at TC Electronic—a Danish company, which develops and manufactures sound equipment for professionals and musicians. TC Electronic has a large product range and special expertise in digital sound processing. TC Electronic's products range from signal processing for studios and computer recordings to digital guitar effects (in the form of digital guitar pedals).

3.5.2.2 Assignment of the Workshop

The assignment in workshop 1 was to create a new or alternative understanding of the context in which TC will market their future products. In order to make the workshop more specific, the focus was narrowed down to consider digital products targeted at guitarists. In this assignment, the objective was also to gain understanding of the insights and perspectives of the users and stakeholders.

3.5.2.3 Participants in the Workshop

The participants in workshop 1 were from both inside and outside TC Electronic. The participants included three professional guitarists, a hardware engineer, a software engineer, a philosopher, a representative from product management and a part-time guitarist, who also worked as a professional sound-tester at TC (Fig. 3.3).

Fig. 3.3 Pictures from the visit to a music-instrument shop, just before the workshop

Fig. 3.4 The Jimi Hendrix Guitarist

3.5.2.4 The Individual Models

The presentation of the individual models revealed that there were two different attitudes towards the sound present in the workshop. (1) The guitarists' models focused on the guitarists' role on stage and the use of monolog sound equipment, and (2) the engineers' models focused on the guitar pedal with digital sound (which they found much more advanced than analog guitar pedals, and therefore more relevant). This difference was, for instance, represented in the model called: 'The Jimi Hendrix Guitarist'.

The Jimi Hendrix Guitarist in Fig. 3.4 was created by one of the professional guitarists. Referring to the guitarist in the center of the stage, who gets all the

Fig. 3.5 The Artistic Guitarist

attention. He is a star, who jumps from band to band—in his search for success. He wants to be like Jimi Hendrix and is very proud of his old-fashioned analog sound equipment.

Another individual model, which was discussed a lot in the workshop and which came to play a significant role in the shared model, was the model called 'The Artistic Guitarist'. It represented a new emerging kind of guitarist, who was different from the Jimi Hendrix guitarist.

The Artistic Guitarist in Fig. 3.5 was created by the part time guitarist and sound tester, who was referring to himself as the new artistic guitarist. This new artistic guitarist is a type of guitarist, who was more artistic in his approach to music—and experimenting more with his use of sounds. He is not the stage person, but the kind of guitarist who is experimenting with his band in the basement. And accordingly, most of his contact to the music industry and the audience happens through the Internet. After the presentation of The Artistic Guitarist, one of the engineers asked if digital sound could be seen as a potential advantage for The Artistic Guitarist due to the possibilities like customization of sound and uploading/downloading new sounds via the Internet. This received a positive response from the part time guitarist/sound tester as well as from the other guitarists, and initiated a further discussion on this.

3.5.2.5 The Shared Model

The creation of the shared model was initiated with some discussion back and forth between digital and analog sound, and the different types of appreciation, which took place. However, due to the time limits of the workshop, participants started to focus on the assignment of building the models together. The outcome of the workshop was the shared model called 'The Cross Road' (Fig. 3.6).

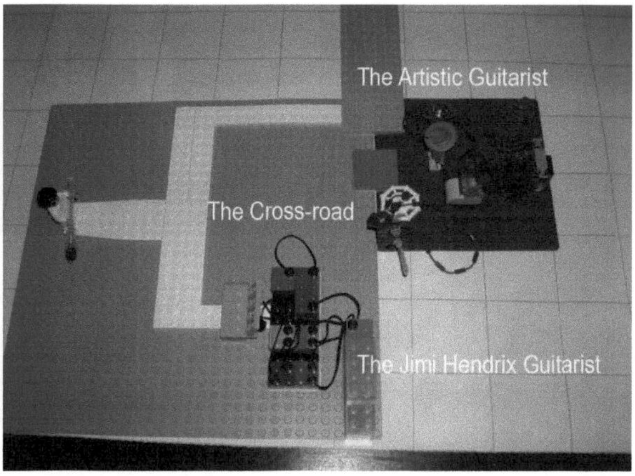

Fig. 3.6 The shared model from workshop 1

It symbolized that TC Electronic had to make a strategic choice between The Jimi Hendrix Guitarist and The Artistic Guitarist as their target user, mainly because their needs and approaches to sound are totally different. As one of the engineers noted: TC Electronic is standing between the two, because they are targeting their products at The Jimi Hendrix Guitarist by using the digital sound technology to imitate monolog sound, instead of developing the potentials of digital sound, as it was suggested in relation to The Artistic Guitarist.

3.5.3 Workshop Case 2: The Base Camp, Red Cross

This workshop was conducted as a Stakeholder workshop.

3.5.3.1 Context of the Workshop

Workshop 2 was held at the emergency aid organization Danish International Red Cross—department for disaster management. The function of the disaster management department is to assist and provide aid in large catastrophes like tsunamis, flooding, or earthquakes. This is done along with Red Cross departments from many other countries.

3.5.3.2 Assignment of the Workshop

In 2005, the Danish department of International Red Cross was given the assignment to provide a base camp for aid workers during their stay in disaster areas. The base camp is for situations of emergent disasters like tsunamis, flooding

Fig. 3.7 Participants in workshop 2

or earthquakes, and it is a temporary installation lasting a maximum of 6 months. At the time when the workshop was held, two situations, in which Red Cross in Denmark had to send out some sort of base camp, had already occurred. Out of necessity, these had been built on the basis of a camp, which Red Cross had inherited from the civil defense. The assignment of the workshop was to initiate the development of the base camp, to gain understanding of the users' insights and perspectives, and to define guidelines on which the development could be based.

3.5.3.3 Participants in the Workshop

Workshop 2 included a number of participants with different insights in relation to the base camp. The participants in the workshop counted a nurse working as a base-camp manager, an emergency coordinator educated in logistics in the military, the emergency chief, an engineer, who had been assembling the first two camps, a voluntary aid worker, and two industrial design students (Fig. 3.7).

3.5.3.4 The Individual Models

The individual models illuminated a variety of views on the base camp—that is everything from purchasing, storing, and packaging the different goods for the camp to arranging and assembling the camp, and also working and living in it. The nurse made an example of this. The model and her explanation showed that she was very concerned about the hygiene and the temperature conditions in the base camp, and she shared insights about how she was often freezing all night, because she did not find it safe to use the burning stove in the tent while sleeping.

Fig. 3.8 The camp model by the engineer

Another example is the engineer's model. This also included the concerns about comfort, but in terms of making the technical system function without spending too much money on it. The engineer's model called 'The Simplified System' is presented in Fig. 3.8 and he explained it as:

> Well, this is a very simple camp (…) We have tried to make a very efficient technical support department. As you see, one of the treasure chests is still full, but the other one is actually empty. The reason for this is that it is very expensive to make all these technical solutions. Therefore, we have to make it as cheap as possible, because otherwise all those who want to get help from the camp end up looking like this [points to the skeletons]. Accordingly, the idea is to make a very simple system (…) without compromising comfort. So that the people out here can be happy (…) that they survived.

3.5.3.5 The Shared Model

When the participants were asked to build their insights together in one model, it resulted in a longer discussion and negotiations about the priorities in the camp. Apparently, it was hard combining the different models.

The final model derived directly from combining the different models. At the end of the workshop, the participants divided the shared model of the base camp into different modules: a module containing the sleeping area, the toilet area, the kitchen area, the working area, and so forth. The model is shown in Fig. 3.9.

As a result of the reflection on the shared model and its influence on the project, an idea about defining certain needs according to the different areas slowly emerged out of the discussion. This idea was also mentioned as prevention against generalized solutions throughout the whole camp—as in the case of the burning stove.

The emergency coordinator, who was in charge of buying and storing the camp, also elaborated on the shared model and the modular structure. He suggested that instead of sending off X numbers of tents and Y numbers of stoves, he would, instead, store the camp in units that were equal to the modules. By doing so, he could more easily adjust to different needs. Furthermore, this type of delivery

Fig. 3.9 The shared model from workshop 2

would make it much easier for the aid workers to assemble the camp. The engineer responded to this by saying that he had told the logistic department about his trouble when assembling the camp over and over again, but nobody had been able to understand him, so far.

3.5.4 Workshop Case 3: The Social System Engineering, Daimler AG

This workshop was conducted as a Team workshop.

3.5.4.1 Context of the Workshop

Workshop 3 took place at Daimler AG (formerly known as Daimler- Benz AG and DaimlerChrysler AG). Daimler AG is a German car corporation, which both develops and manufactures automobiles and trucks. In addition to the development of vehicles, Daimler also provides financial services. Daimler is a very large organization, which includes many different professions, among this a group of industrial psychologists positioned in Data and Process Management in Ulm. It was in this department workshop 3 was held.

3.5.4.2 Assignment of the Workshop

A project called Social System Engineering Program was positioned in workshop 3, in the very beginning. The aim of the Social System Engineering Program was to provide tools and methods to engineering teams in development projects, in order to optimize their collaboration. The workshop objective was mainly to assure a shared understanding of the project within the team.

Fig. 3.10 Participants in workshop 3

3.5.4.3 Participants in the Workshop

The participants in workshop 3 were all from the project team. They were all employed in the same department, and each one had a background in industrial psychology. Out of all the workshops presented in this book, workshop 3 included the smallest and most homogeneous group (Fig. 3.10).

3.5.4.4 The Individual Models

Although close colleagues created the models in workshop 3, there were two different kinds of models. On the one hand, the models were viewing the Social System Engineering Program as a linear process, where the psychologists were to first observe the collaboration in the engineering projects, diagnose the situation and then make a number of interventions, by introducing different tools and methods—and then finally arrive at an optimized situation.

One of these models was referred to as 'The Linear Approach' shown in Fig. 3.11 and was explained in this way:

> These are the situations [that we in the industrial psychology group are trying to approach] – [There are] different kinds of projects in the company – and they are complex in different ways (...)Then we are doing a diagnostic. (...) we see the different kinds of problems, the different kinds of people. We see politicians, we see enemies and problems. (...) and we have the intervention tools for the situation. (...) [And after that the team] now knows how to work on it and they now have the knowledge. And everything is structured for them – and they can work on their aim. And we just see it works and we observe it – then we are free (...)

On the other hand, one of the models presented Daimler AG as an organization with many different behavioral levels. The model showed that even if the

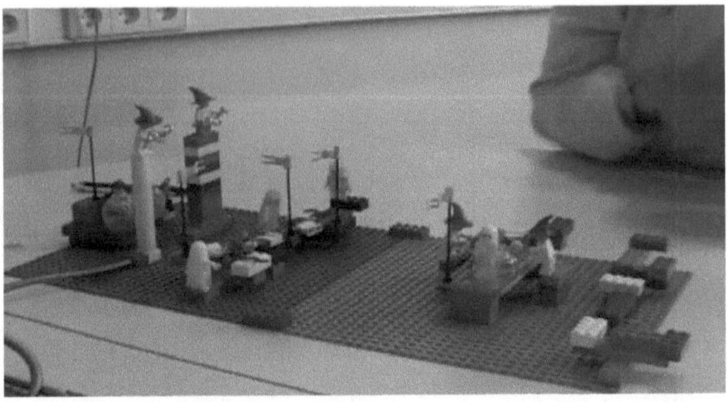

Fig. 3.11 The linear approach

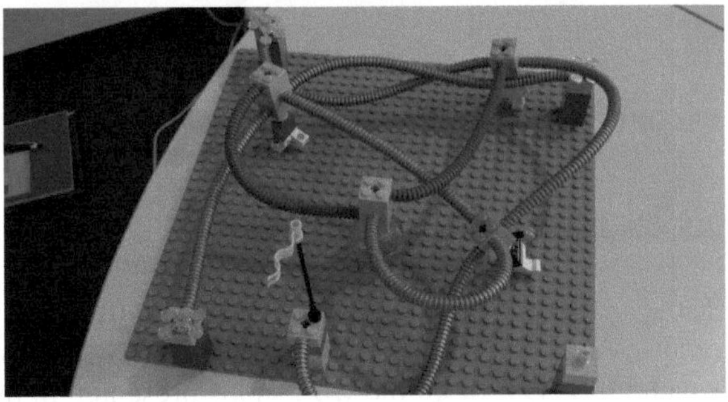

Fig. 3.12 Running Away from the Task

development teams were put together to do an assignment, there could be many different reasons why this did not happen. For instance, personal wishes, directions, or even counter orders from the home department.

The model is shown in Fig. 3.12 and was described with these comments:

> Here are three people sharing one task, which they have to collaborate on. But in the organization there are so many different behavioral levels influencing them and driving them into certain directions. That is also connected with their motives, their plans, their goals (…) And this is also [influenced by]: 'Where can I get money?'; 'What behavior am I rewarded for?' (…) And so in the end these three people – because these behavior steering structures show different directions (…) they are completely lost and running away from the task [symbolized by the blue sphere in the middle of the model] – instead of meeting in the middle of it and working on it.

3.5.4.5 The Shared Model

When the team was given the assignment to build the models together, it caused some discussion. The discussion was mainly focused on whether to focus the chaotic description of the development teams (represented in the model, which they gave the name 'Running Away from the Task') or on the idea of diagnosis, intervention, and evaluation (represented in The Linear Approach).

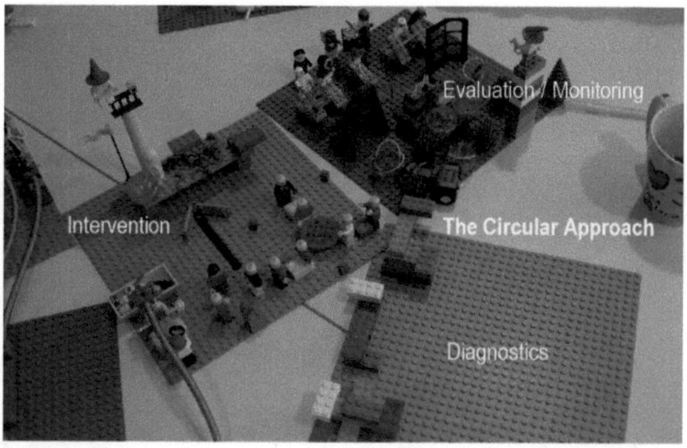

Fig. 3.13 The shared model from workshop 3

However, the shared model (Fig. 3.13) ended up combining both the insights from Running Away from the Task and the Linear Approach. The result was a model called The Circular Approach. This model viewed the Social System Engineering Program as a circular process, where the psychologists would constantly be monitoring the collaboration, making diagnosis, and interventions. This also included the understanding of the Social System Engineering Program as an ongoing task or assignment, not as an accomplishment, or time limited project.

3.5.5 Workshop Case 4: The Medical Treatment Houses, Region Northern Jutland

This workshop was conducted as a Stakeholder workshop.

3.5.5.1 Context of the Workshop

Medical Treatment Houses is a project initiated by Region Northern Jutland and the design firm Smedegaard and Weis, with financial support from The Danish Enterprise, and Construction Authority. The idea of shared Medical Treatment Houses, which can include doctors, nurses, physiotherapists, and other health services, is a response to the shortage of medical practitioners in the northern part of Denmark. The aim of the project is to change the statistics showing that within a few years, the number of citizens without a medical practitioner within a distance of 15 km will increase dramatically. The idea of Medical Treatment Houses is especially going to motivate and attract young female doctors (most doctors in Denmark will be females in future), since it can provide flexibility in relation to working hours and further possibilities in relation to vacation, maternity, and parental leave.

Fig. 3.14 The participants from workshop 4

3.5.5.2 Assignment of the Workshop

The aim of the workshop is to find a set of guidelines or a model for developing these Medical Treatment Houses.

3.5.5.3 Participants in the Workshop

The participants of the workshop 4 were: From Region Northern Jutland: the coordinator of the project as well as the director of the medical practitioner sector. From practice: one nurse, one secretary and one medical practitioner, who work in one of the first Medical Treatment Houses in Denmark. And finally from the area of design: two designers from Smedegaard and Weis, along with an Associate Professor in Design from Aalborg University (Fig. 3.14).

3.5.5.4 The Individual Models

In the beginning of the workshop, there were many different models and therefore also many different perspectives on the project. In the model (Fig. 3.15) by the director from Region Northern Jutland, the aim to solve the problem of shortage of medical practitioners was very apparent and explained like this:

> Today, the structure for medical practitioners means that we have a doctor in every little town. And these doctors each have a set of patients. But [this doctor] he will soon be retiring – and this means that there will not be a new one. Because this [female doctor] she does not want to go there and work. And this means that the patient in the town has a problem. That is why we need to find a place, where she wants to work, and she wants to work in a treatment house (...) This is the assignment - as I see it.

Fig. 3.15 The perception of the current situation by the Director of Region Northern Jutland

Fig. 3.16 Model by nurse

The nurse's, secretary's and the medical practitioner's models were very concerned with the physical layout of the Medical Treatment Houses. An example of this is the secretary's model in Fig. 3.16. which she commented with:

> I am imagining a light and friendly house with many windows and good air-conditioning. (…) When you come into the house you will immediately meet a person – in case you have any questions. (…) I would also like a second floor, where the staff can stay between shifts (…) and maybe some outdoor areas.

The designers' and project coordinators' models were generally focused on the guidelines or key issues to be approached in order to make the Medical Treatment Houses function. Issues such as flexibility, to make sure that the Medical Treatment Houses are attractive and inviting enough for the medical practitioners to invest in it. (In Denmark medical practitioners work as self-employed business owners—even though most of their salary comes directly from the government). Issues like 'social glue' or 'fellowship', which can drive the development and collaboration in the Treatment Houses, were also important. The model called 'The Community Mill' is an example of a model of this nature.

Fig. 3.17 The model pulling together

3.5.5.5 The Shared Model

When building the individual models together into a shared model, there was a tendency just to gather the individual models in the middle of the table rather than negotiating which models were important. Therefore, the group was also asked to build a set of individual models, which showed most important influences, in relation to the success of the Medical Treatment Houses and to rate these subsequently.

One of the models with the highest ratings was the model called: 'Pulling Together' (or directly translated from Danish: Lifting together) It was created by the director from Region Northern Jutland and is shown in Fig. 3.17. He explained the model like this:

> Economy is a very important factor in relation to the establishment of these houses. But before that, the definition of a need emerges and the definition of a wish to do this. And [when this is settled] it is necessary that some different parties are pulling together. This is what [the model] symbolizes. (…) There are the medical practitioners and the local authorities – a municipality and a region, which are very relevant parties – in the process of making this happen (…) These parties must agree on the project in order to make it happen and they must agree to pull together.

The other model, the Community Mill (Fig. 3.18), was explained like this:

> (..) I have not divided [my model of the Treatment House] into rooms. Instead I have created this [mill], which is the focal point in the community. It can be shared facilities, but at the same time something that keeps the house running [socially]. I have also created a minor zoo (…) It symbolizes that there should be room for diversity. I imagine different types and professionals, who are able to share accommodation (…).

During the reflections after the workshop, The Community Mill and the model called Pulling Together were discussed as some of the most important guidelines in relation to the Medical Treatment House. At some point in the discussion they were even merged together in terms of the idea of Pulling Together, at all levels and at all times. From a specific Medical Treatment House, where the region, the

Fig. 3.18 The model community mill

municipality and a number of practitioners have to pull together their interests, and invest together, to the everyday operation of a Medical Treatment House, which requires that different people working there have to pull together to make the place function for the citizens and themselves.

3.5.6 Workshop Case 5: The Good Elderly Life Part 1, Copenhagen Living Lab

This workshop was conducted as a Team workshop.

3.5.6.1 Context of the Workshop

Workshop 5 was the first of the two workshops in The Good Elderly Life project. In the initial description of The Good Elderly Life project, the focus of the project or the frame for improvement was the nursing home. According to the initial project application the project objective was:

> (…) to gain insight into what 'the good elderly life' is, when living in a nursing home and identifying the innovation potentials within elderly care. [2]

3.5.6.2 Assignment of the Workshop

Workshop 5 was placed in the beginning of The Good Elderly Life project with the intention of ensuring an internal clarification and creating a shared understanding of the project within the interdisciplinary team and the many different partners.

Fig. 3.19 The participants in workshop 5

3.5.6.3 Participants in the Workshop

The participants in workshop 5 included two project developers from Copenhagen Living Lab, an ethnographer doing research in the nursing home, two project managers from the Health and Care administration as well as two deputy managers from the nursing home Sølund (both originally educated as nurses) (Fig. 3.19).

3.5.6.4 The Individual Models

In the first part of the workshop, the models could be divided into two groups. Some of the models stayed within the nursing home context, while others attempted to look at the elderly life more in general. It was especially the participants from the nursing home and the municipality, who stayed within the initial description. An example of this is the model made by the deputy manager Lene who said (Fig. 3.20):

> My point of departure is my job and here you find several different interests. There are those [elderly people] who prefer to be active. (…) But there are also those who think: 'this is enough – let's sit down and relax in our little house' (…) 'it is far too much (…) It is not necessary to have activities all the time'. It symbolizes that different people want different things.

One of the participants, who built a model related to the elderly life in general, was the project manager Thomas. His model called 'The Good Life' viewed the elderly life not as something separate from, but as part of life.

Thomas explained his model, illustrated in Fig. 3.21:

> (…) The Good Elderly Life – is here [on the right]. However it is not different from the rest of my life, and the rest of The Good Life (…) you are not different because you are old –you are just in another period of your life. And all the things, which you have

Fig. 3.20 Model by Lene; people want different things

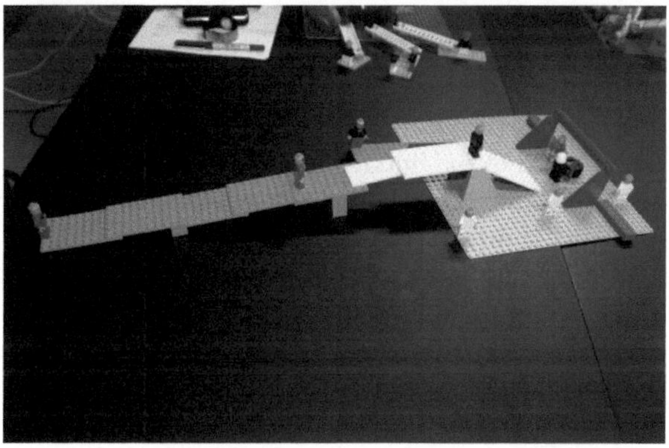

Fig. 3.21 The good elderly life

experienced (…) are still part of you. [In the elderly life] there is a need for protection, which has not been there earlier – it is more flat, where you live the last part of your life, but there is still experiences. (…) And there are connections back to the life, I have lived – my family, my relatives, who are still part of my life.

Apart from these two directions there was also a number of models, which included the political and organizational system surrounding the nursing home and The Good Elderly life.

3.5.6.5 The Shared Model

When participants began to build the different models together, everyone agreed to start with 'The Good Life' model and then from this build the Good Elderly Life. However, the last part of this plan created some discussions about how to organize the Good Elderly Life.

In the beginning, the Good Elderly Life was mainly put together by the different models, which showed the elderly life in the nursing home. However, at some

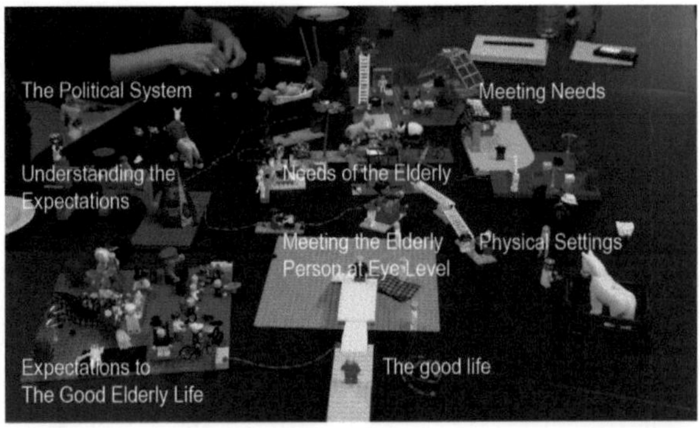

Fig. 3.22 The shared model from workshop 5

point, it was suggested that instead of looking at the physical settings, it would be more important to look at the needs of the elderly people.

The needs of the elderly person became the focus of the shared model, and the understanding of the contexts outside the nursing home frame became a leading idea in the shared model. This meant also that several new and additional models were added to the initial shared model, like a model that showed expectations and preconceptions in relation to how the life of elderly people unfolds. The shared model is shown in Fig. 3.22.

During the reflections on workshop 5, the impact of the shared model was discussed in relation to the project, and it was found that an understanding of The Good Elderly Life could not be gained only through investigations within the present nursing home framing. Instead, it required a much more holistic approach. In the time after the workshop this change in understanding also affected the activities in the project, i.e., the ethnologist's work.

A few weeks after the workshop, project manager Thomas Hammer- Jakobsen from Copenhagen Living Lab summed up workshop 5 by saying that:

> We realized that we had to look at needs, experiences and understandings in a broader scope. It is harder than the initial intentions, but it is also important. Otherwise we will just create a reproduction of the construction, which lies within Sølund's [the nursing homes] bricks. [1].

3.5.7 Workshop Case 6: The Good Elderly Life Part 2, Copenhagen Living Lab

This workshop was conducted as a Stakeholder workshop.

3.5.7.1 Context of the Workshop

Workshop 6 took place 4 months after workshop 5. At this time in the project, there was a lot of discussion and enthusiasm in the team about assistive technology and its potentials in relation to elderly people.

Fig. 3.23 Pictures from the workshop 6 in the Good Elderly Life project

3.5.7.2 Assignment of the Workshop

Workshop 6 was implemented as a means to create dialog with various stakeholders to encourage their involvement in the project and to incorporate their insights and perspectives.

3.5.7.3 Participants in the Workshop

Besides the interdisciplinary team, Workshop 6 also included elderly people, nursing home residents, relatives, and employees in nursing homes. The participants can be seen in the pictures in Fig. 3.23.

3.5.7.4 The Individual Models

In workshop 6 there were many different individual models, which became part of the shared model. In the section below many of them will be presented. The first individual model, which will be presented here, is called The Gate to The Elderly Life. 76-year-old Eva, who is a resident at the nursing home Sølund, built it. The gate symbolizes the realization and acceptance of being old and voices both frustration and difficulties in relation to the physical and mental situation.

Eva explained her model this way, illustrated in Fig. 3.24:

> (…) I may well say this Gate – it is a bit difficult to enter. You have to acknowledge that you are old and that there are many things you cannot do. Some of the obstacles are higher than others. That is what I think being old - is like. You feel the limitations in many ways. (…) You also think differently. You are more easily (…) confused.

Fig. 3.24 The gate by Eva

Fig. 3.25 The model Brown mass by Josefine

Eva also expressed a need for help to handle this situation both from relatives and from the nursing assistants. She also said that at the moment, no one seems to address this issue. During one of the workshop breaks, she further elaborated on the importance of the gate by saying that if you do not enter that gate, it does not matter how well the meals are prepared, because if you have not accepted that you are not able to cook your own meals, everything will taste bad. Similarly, if you do not accept that you need help, all help will be inadequate.

Another individual model from workshop 6 was the 'The brown mass' shown in Fig. 3.25. The ethnologist Josefine built it, on the basis of her first and preliminary studies of the nursing home context. She explained the model like this:

> (…) the problematic part about [being old] may well be the emergence of (…) an overall societal opinion (…) You are easily detached from society. You are seen as some kind of brown mass(…) you are trapped in your own world (…) maybe because you forget who you used to be (…) And being in a nursing home (…) it is just storage on your way to death – you are not getting home again. It is the last place (…)

Josefine's model contains a number of things, from the importance of life stories in dialog with elderly people, to the acknowledgment that phrases like the

Fig. 3.26 The model 'To give and to get help'

Fig. 3.27 The human rigths model

'the burden of the elderly' might be harder for 'the burden' than for the ones carrying it, and finally that the confrontation with death can be rather depressing.

A third important individual model from the workshop was called: 'To give help and to get help' shown in Fig. 3.26. This model acknowledged that getting help without being able to give something back may feel degrading or even humiliating:

> I have a partner over here, who does not help me because he is supposed to, but because we are in the same situation, and because he wants to (…) we have some agreements on giving help and not so much about getting help. (…)

The 'To give help and to get help' model also rethinks the nursing home situation into a place, in which everyone plays an active role, even if it means peeling one carrot a day or reading a small section in the newspaper to someone.

The last example, which will be provided here, is the 'human rights' model shown in Fig. 3.27. It is concerned with the rights, which elderly people have or should have, for instance, being treated with respect and being accepted for one's individual needs, wants, and habits. The 'human rights' model was built by the nursing assistant Pia, who explains:

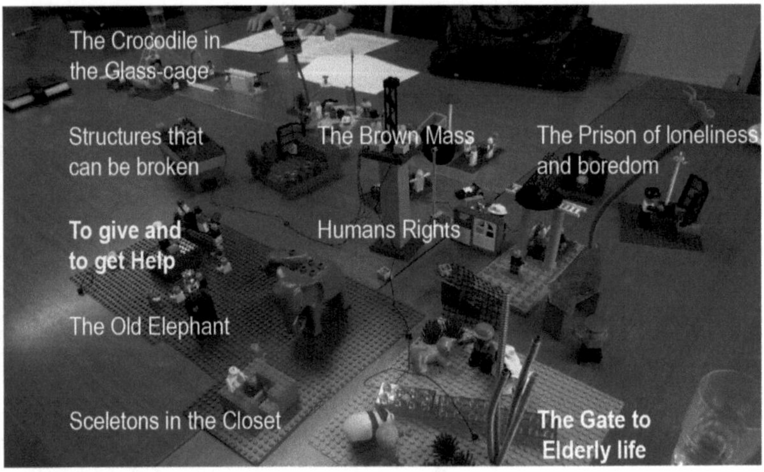

Fig. 3.28 The shared model from workshop 6

> I somehow had human rights in the back of my mind. (…) [It is about] being able to go to the toilet, when needed and not at a certain time. (…) If you prefer a shower in the evening – yes, why not? – Why does it have to take place in the morning? I hope that it will be further acknowledged that we are different and that we have different habits and needs and that resources will be distributed differently than they are today.

3.5.7.5 The Shared Model

The shared model in workshop 6 was actually a collection of the most significant individual models, models that everyone agreed were important. The models included insights on both functional, personal and social needs, and experiences. In the Fig. 3.28, there is a picture of the final outcome.

In the evaluation of workshop 6, Thomas Hammer-Jakobsen called the Lego models a first step towards 'design principles' or 'design dogmas', which could be used in the development of the design briefs. The idea was that all briefs in the project should include or consider: 'to give help and to get help', 'human rights', 'the gate to the elderly life', and so forth.

He also saw the outcome of the workshop as a form of evaluation scheme, which could be used when evaluating future concepts in the project. The creation of the Lego models also proved to have a further impact on The Good Elderly Life project. In the ethnological fieldwork following workshop 6, the ethnologist was able to identify and underpin the outcome of the workshop, and, for instance, find evidence of problems like The Gate to the Elderly life and the need To Give and To Get Help. Furthermore, she chose to use these 'one-liners' in the presentation of her analysis, because, as she argued, they each held several senses and experiences, which otherwise would be hard to combine into meanings.

When the ethnologist presented the fieldwork and analysis to a panel, which included employees both from Copenhagen's Health and Care administration and from several care facilities, it was concluded that many insights on elderly people were recognizable to the group. However, as one of the participants noted, because the insights were gathered in this particular way, they created a whole new understanding—and new perspectives on how to handle them.

3.6 Summary

In this chapter, the workshop method and the research material have been presented. Firstly, the workshop background, set-up, and development were reviewed in terms of LSP.

Secondly, the workshop portfolio was presented. In this section, there was a brief introduction to the four single workshops and to the two workshops in the longitudinal study. Further, the characteristics of the workshop portfolio were presented in order to underline the broad variety of projects and organizational contexts, and the quality of the data, in terms of being real time, and real projects. And finally, each workshop was presented in terms of context, type, assignment, participants, and the process from individual to shared model.

References

1. Copenhagen Living Lab (2008) Meeting minutes, evaluation meeting I, 24 April 2008
2. Danish Business Authority (2007) Application for funding/Det gode ældreliv. http://www.ebst.dk/brugerdreveninnovation.dk/projekterbrugerdreveninnovation/0/14/5048636
3. Gauntlett D (2007) Creative explorations—new approaches to identities and audiences, Routledge, London
4. Gruber H, Vonéche J (1977) The essential piaget, Basic Books, New York
5. Hey JHG, Joyce CK, Beckman SL (2007) Framing innovation: negotiating shared frames during early design phases. J Des Res 6(1–2):79–99
6. Lego Serious Play (2006) The science of Lego serious play. Lego, Billund
7. Papert S (1993) The children's machine—Rethinking school in the age of the computer, Basic Books, New York
8. Schrage M (2000) Serious play—how the world's best companies simulate innovation, Harvard Business School Press, Boston
9. Terr L (2000) Beyond love and work—why adults need to play. Touchstone, New York

Chapter 4
Identifying General Patterns

In this part of the book, the analysis of the research material and main findings of this study will be presented. In the following chapters, the aim is to identify patterns across the six workshops and to show the main findings. The analyses and findings will be divided into three chapters. In this chapter, there will be a presentation of the general patterns across the six workshops. This will narrow down the unit of analysis to a set of significant models, which had an important influence on the workshops. Chapter 5 holds a detailed analysis of the significant models and provides examples of their influences in the workshop. Furthermore, it identifies the reasons behind these influences, based on the insights from the theoretical framework. In Chap. 6, the significant models will be compared to existing definitions. It will be qualified why these significant models are to be seen as 'new' and not previously described, and their names and definitions will be presented.

4.1 General Patterns

When analyzing the video material from the workshops and the additional documents from the longitudinal study, it was evident that some of the Lego models played a more significant role in the workshops than other models. These models were often created by some of the participants with direct experiences from the context of use, and they enabled the stakeholders to make some of their sticky knowledge explicit. Another characteristic of these models was that they provided a second-order understanding of their creators' meaning making and thereby supported the communication in the group. The significant Lego models were also an important part of the shared models, which were created in the workshops. In relation to the shared models, the significant models enabled the participants to create a shared frame in relation to the project. When analyzing the significant Lego models more closely, it was also evident that their structure was different from the other Lego models created in the workshops. Whereas the ordinary

L. Møller and C. Tollestrup, *Creating Shared Understanding in Product Development Teams*, DOI: 10.1007/978-1-4471-4180-8_4, © Springer-Verlag London 2013

Fig. 4.1 The model "To Give and To Get Help"

models either represented a concept or an experience/reflection, the significant models had both a 'concept component' and an 'experience component'.

An example of this can be found in workshop 6 and the Good Elderly Life project. In this workshop, there were five significant Lego models, including the model called: To Give and To Get Help shown in Fig. 4.1.

> I have a partner over here, who helps me not because he is supposed to, but because we are in the same situation, and because he wants to (…)We have some agreements on giving help and not so much about getting help. (…)

The 'concept component' in this significant model is 'to give and to get help' and the 'experience component' is the creator's personal story about how humiliating and maybe even dehumanizing it is, if you are in a situation, where you cannot give anything back. The concept 'to give and to get help' can be seen as the headline or perspective in which all creators' experiences can be linked and interpreted. The 'experience component' is the personal experience, which links the significant model to its creator and to a concrete set of situations.

Another way of explaining the difference between a regular model in the workshop and the significant models is by comparing the significant model called: To Give and To Get Help to one of the regular models from workshop 6 shown in Fig. 4.2.

The regular model expressed a concept and was explained like this:

> I have chosen to look at it from a technological perspective – and I hope we will be able to reach this within 10 years. I have tried to illustrate it with some kind of station (…) combined with a private robot, which can help me [the elderly person] with the things I want and need. (…) If it was inside a nursing home, every room would have its own robot to wait on [the elderly person] hand and foot – and anywhere else it is wanted (…)

When comparing the regular model to the significant model 'To Give and to Get Help' it becomes clear that the 'experience component' is missing in the regular model; there are no insights on whether the creator would like a robot or not—or what kind of experiences or motivators are behind these suggestions or

Fig. 4.2 The regular model from workshop 6

ideas about a robot.[1] The concept in the regular model represents an idea, where as 'To Give and to get Help' represents a concept of how its creator makes meaning of the nursing home situation. In Fig. 4.3, there is an illustration of the difference between the regular model (on the left) and the significant Lego model (on the right).

As the figure illustrates, the significant Lego model is not detached from the person, who creates it. Instead, it holds both the 'concept component', which represents how this person makes meaning, and the personal experiences, which are the background for this meaning making–shared in terms of a storytelling.

Another difference between regular models in the workshops and the significant models was that the significant models represented a way of understanding all the experiences each stakeholder or team member had been gathering. In the example of the 'The Gate to The Elderly Life' (also from workshop 6), 76-year-old Eva combined a number of experiences and emotions from her own life and transformed these into the significant model of the 'Gate' shown in Fig. 4.4.:

> (…) I may well say this Gate – it is a bit difficult to enter. You have to acknowledge that you are old and that there are many things you cannot do. Some of the obstacles are higher than others. That is what I think - being old - is like. You feel the limitations in many ways. (…) You also think differently. You are more easily (…) confused.

This 'Gate' made it possible for her to overcome the boundary of not being able to express her opinions and values to the project team. This was possible due to the fact that the model contained a lot of experiences, which otherwise would be hard to connect into something meaningful for somebody, who had not had those personal experiences.

The experiences in relation to The Gate might have been: 'I am not able to do, what I used to do—and it makes me sad', 'I experience the limits of my physical decay everyday", "I do not like the food here", "I feel as if nobody understands me' etc. Expressions like these might have been difficult to relate to for the other

[1] Facilitation Guide Sect. 8.2.3: What to listen for in team discussions.

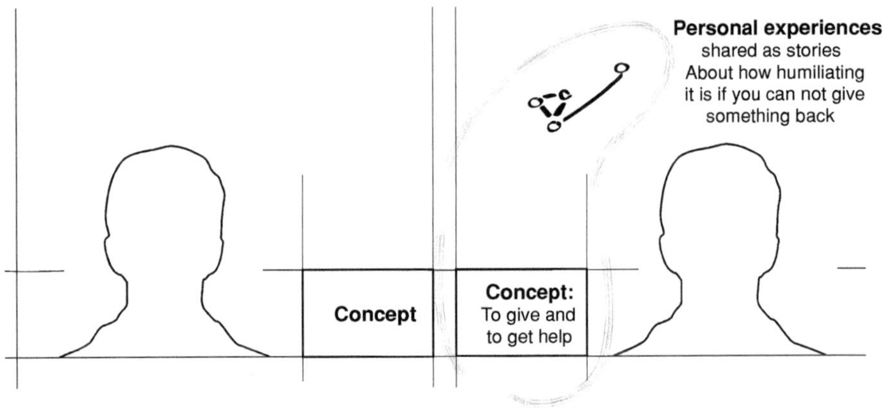

Fig. 4.3 The difference between a regular model and a significant model

Fig. 4.4 The gate model

participants in the workshop, but with The Gate to the Elderly Life, they make sense. And just as important, the other stakeholders or members of the team can also add similar experiences to the 'concept' in their personal meaning-making process. In Fig. 4.5 there is a simple illustration of this. In the illustration, all the little black dots represent experiences. If the experiences are shared via a story or in a model (as viewed on the left), it can be hard to interpret how these different experiences are connected, and thereby also what is meaningful to the person, who is sharing it. However, the significant Lego model (reviewed on the right in Fig. 4.5) reveals how these connections are made via the 'concept component'.

Other stakeholders or members of the design team can also add similar feelings and experiences to 'The Gate' in their personal meaning-making process. For instance, if you have an older relative, who is complaining about the food in the nursing home, or who is depressed about not being able to clean his or her house properly, these experiences can perhaps be better understood in the light of The Gate to the Elderly Life. Another way to illuminate this is by showing one of the

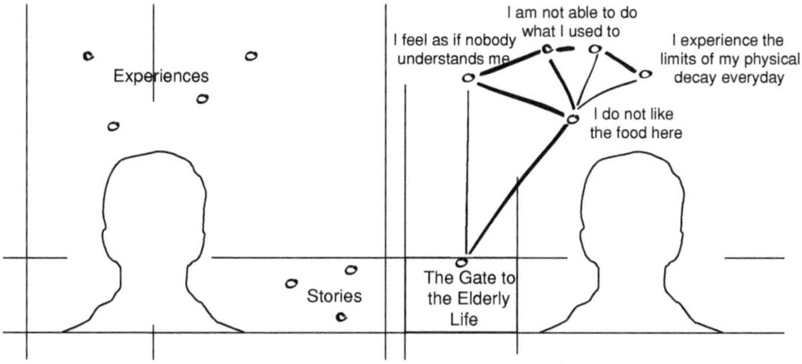

Fig. 4.5 The difference between sharing experiences and a significant model

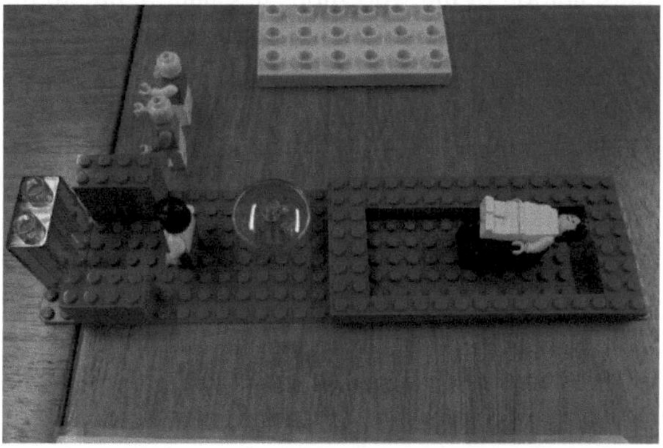

Fig. 4.6 The regular model from workshop 6

regular models from workshop 6, which only included an experience/reflection (or the 'experience component').

The regular model shown in Fig. 4.6 was created by a relative to one of the elderly people living in the nursing home, and she explained the model as follows:

> (…) Well… I can of course only talk about my mother, who is 90 years old and lives at Sølund [the nursing home], where I am really happy she is. And she is happy as well—however sometimes is it a little bit difficult (…) because even though she can hardly do anything, she does not like to be of any inconvenience to the staff. But then of course she has me. (…) I often come around, and we eat together and spend time together. I also wash for her and she likes that (…) she still likes to be neat (…) [She is not that social any more] She often just sits and stares into space, because she does not like to listen to tapes any more and she cannot watch television any more, and then life becomes a little bit sad. There are many depressions.

As it appears, the relative shared a lot of personal stories and insights in relation to her mother. However, it can be rather hard to combine these experiences and understand how she makes sense of these.

Table 4.1 The Lego models from the six workshops divided into reflections, concepts and significant Lego models

Workshop	1[a]	2	3	4	5	6
Concepts	–	4	6	5	6	5
Experiences/Reflections	–	10	2	7	5	8
Significant Lego Models	2	1	4	2	4	7

[a] Since the TC—Electronic workshop was not video-documented, there are not exact accounts of the reflections, concepts and significant models. However, due to the pictures and notes from the workshop, it is certain that there are two significant models.

4.2 The Significant Models

A large number of the significant Lego models have already been shown in the presentation of the workshops in Chap. 3, due to their influence on the workshops and the workshop projects. The significant models are, for instance: The Artistic Guitarist from the workshop at TC Electronic, The Modular Structure from the Red Cross workshop, Running Away From the Task from the workshop at Daimler, The Community Mill in the workshop about Medical Treatment Houses as well as The Good Life and The Gate to the Elderly life, from the Good Elderly Life workshops.

However, far from all the models created in the workshops turned out to be significant models. Typically, the significant models:

1. Enabled the stakeholders to make some of their sticky knowledge explicit.
2. Provided a second-order understanding of their creators' meaning making.
3. Supported the communication in the group.

And, as opposed to all the other regular models (concepts, experiences/reflections), the significant models had both a 'concept component' and an 'experience component'.

In Table 4.1 all the Lego models from the workshops are categorized in terms of concepts, experiences/reflect1ions or significant models. As the table shows, the significant models are only a small part of the complete number of models.

In the six workshops, a total of 78 models were created, out of which 20 were significant models.

However, even if the significant models were small in number, they had an important influence on the workshop and are therefore also the units of analysis or research objects of the book.

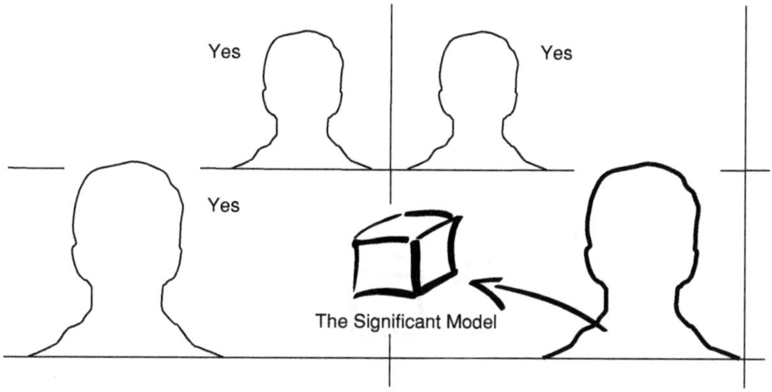

Fig. 4.7 Ownership via acceptance

4.3 From Personal to Shared Significant Models

Most of the significant models were created by an individual and then taken into the shared model. However, there were also alternative ways for the significant Lego models to be created and accepted into the shared model. In the section below, a review follows related to the movements of the significant Lego models into the shared Model, including:

- Adopting the significant personal models into a shared model.
- Combining two individual personal models into a shared model.
- Combining a 'concept component' and an 'experience component' into a shared model.

4.3.1 Adopting the Personal Significant Models into the Shared Model

In the workshops, there were several examples of significant personal models being taken directly into the shared model. This happened when the other participants started to invest ownership in these models, either by stating their personal acceptance of the model or by adding similar or related experiences to the model—and thereby making it shared. This is illustrated in Fig. 4.7.

An example of this can be found in workshop 5, where a number of the participants stated their liking of a personal significant model, created by one of the workshop participants and thereby made it part of the shared model:

Frida: I really like Thomas' model
Lise: yeah, yeah….
Lene: Yes, that one, that is really important. I think it has to be part of it [shared model].

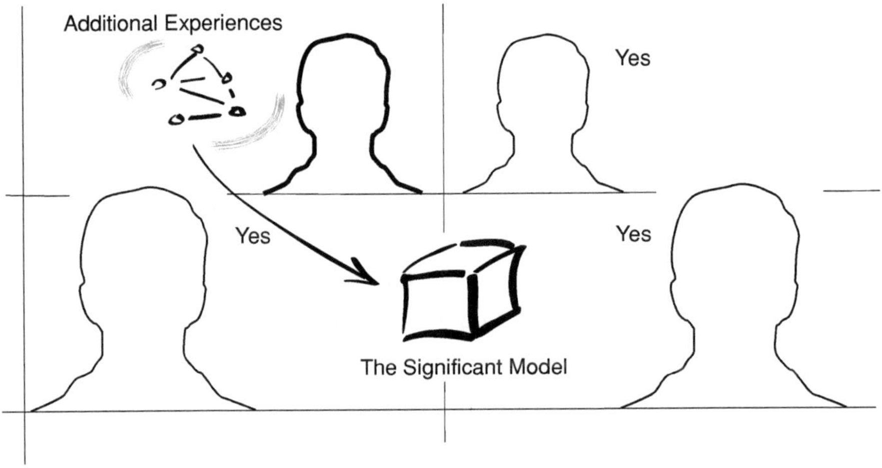

Fig. 4.8 Ownership by sharing additional experiences

In workshop 6, one of the participants showed her ownership in relation to The Gate to the Elderly Life by adding her personal experiences to it (see figure 4.8):

> Lene: (…) But it could also symbolize other things (…) Are you afraid to go through [the gate] because you think it means that someone will start controlling your life? (…) I have experienced that [fear] many times - Especially when we had the protected residences [at the nursing home]. Their greatest fear was to lose their independence. (…)

A similar example to this can be found in workshop 1 at TC Electronic. After the guitarist/sound tester's presentation of The Artistic Guitarist, some of the engineers started to share some of their insights and experiences on digital sound by sharing examples of how The Artistic Guitarist would be able to customize and create new sounds via the digital sound technology, and then share it with his colleagues via the Internet.

4.3.2 Combining two Individual Significant Models into a Shared Model

Sometimes, the process of creating the shared model was not just a matter of accepting and showing ownership in relation to the other participants' significant models. In some of the workshops, two significant individual models were combined into a new one.

One of the most obvious examples of this can be found in workshop 3 at Daimler. In the beginning of the workshop, there were two significant models: The Linear Approach and Running Away from the Task.

In the process of building the models together, The Linear Approach and Running Away from the Task were combined into a circular approach, where the idea

Fig. 4.9 The circular approach from workshop 3 at Daimler

Fig. 4.10 Model of ethnologist's experiences/reflections

was to constantly diagnose, make interventions, and evaluate. The Circular Approach combined both the experiences of the many 'behavioural drivers', which constantly challenge the collaboration in the engineering teams, as well as the approach and experiences linked to industrial psychology. This is shown in Fig. 4.9.

4.3.3 Combining a 'Concept Component' and an 'Experience Component' into a Shared Model

Still, the shared models were not always based on personal significant models only. Sometimes, the process of creating a shared model also involved the combination of a new concept, along with some of the previous experiences/reflections presented in the models, in other words combining a 'concept component' with an 'experience component'. An example of this can be found in workshop 5, shown in Fig. 4.10. In the first part of the workshop, the ethnologist Josefine shared some of her experiences and reflections from her nursing home research:

Fig. 4.11 The *green* and *gray* building bases are used to collect the different experiences/reflections in relation to the different parts of the Red Cross Base Camp

Josefine: (…) My point of departure is: What is the good elderly life for the elderly person? And I have tried to illustrate this as a pictogram. (…) Well it is about being physically active [points to the man on the skateboard]. (…) To have all one's physical faculties [points to the magnifying glass]. (…)To be able to go out and explore. [points to the backpack, camera and flippers] (…) And if you need it, to have the necessary assistive tools [points to the man with the sack trolley as a symbol of a walker] (…) and not to be afraid of new things (…) and to be able to communicate with one's relatives as communication is done today [points to the computer]. (…) To be mobile and to be somewhat financially independent (…)

Later in the process, one of the other participants initiated the creation of a concept or perspective, around which Josefine's observations and experiences could be connected and understood:

Thomas: Well… the way I understand Josefine's model – is that it is almost inside the elderly person. That we are almost down to what is important for the individual.

And after further discussions, Josefine ended up creating the concept herself by stating:

Josefine: Maybe it is 'The expectations towards the elderly life'

By doing so, Josefine added an overall concept or perspective that linked together her experiences/reflections. In relation to the workshop, she also transformed her initial regular model into one of the significant models.

Another example of this can be found in workshop 2 at Red Cross (Fig. 4.11). In the process of creating a shared model of the Base Camp, the participants used a number of LEGO building bases to collect the different experiences/reflections in relation to the different parts of the camp.

At some point in the process, the facilitator interrupts this building process, and asks the participants some questions in relation to what they have created

Poul (facilitator): (…) Well… some of the things I find interesting here is the modular-structure, which you have created now. Is that a way of thinking, which you usually use? (…)
Birgitte: I don't know – H.C. – maybe you can answer this?

H.C: Well…… it is probably the way I look at it. If you isolate every part here (…) then you are able to see what is essential for each module. And then afterwards you can see what will be essential in order to link this one and that one. (…) In this way you get a quick overview of the critical issues, and it also gives you a much better understanding of how [the base camp] can be set up.

As the discussion from the workshop above reveals, the participants have created a concept ('The Modular Structure'), in which all their experiences/reflections can be understood and combined. However, the value or influence of this significant model was being created without anyone being fully conscious of it.

4.4 Summary

In this chapter, the general patterns across the six workshops have been presented. It was found that some of the Lego models played a more significant role in the workshops than other models. And that these models influenced both the communication of meaning and the creation of shared frames in the workshops.

When analyzing the significant Lego models more closely, it was further found that their structure was different from the other Lego models, because they had both a 'concept component' and an 'experience component'.

The number of significant models across the workshops was also presented along with several examples from the workshops. In the last part of the chapter, the significant models' transformation from individual models to shared models was in focus. It was found that not all the models were individual significant models, before they became shared significant models. Instead, it was found that a shared significant model could be created in three ways, either by:

1. Adopting the significant individual models into a shared model.
2. Combining two individual significant models into a shared model.
3. Combining a 'concept component' and an 'experience component' into a shared model.

Chapter 5
The Significant Models

In the beginning of Chap. 4, a number of findings in relation to the significant models were presented. It was reviewed that the significant models (1) Enabled the stakeholders to make some of their sticky knowledge explicit. (2) Provided a second-order understanding of their creators' meaning making, and thereby supported the communication in the group. (3) And that the shared significant models enabled participants to create a shared frame in relation to the project. In this chapter, a number of examples to unfold these statements will be reviewed as well as an explanation of why the significant models were able to accomplish this.

5.1 Making Sticky Knowledge Explicit

One of the best models, to exemplify that the significant models enabled participants to make some of their sticky knowledge explicit, is The Artistic Guitarist created in workshop 1 at TC Electronic. A part time guitarist and sound tester working at TC Electronic created the Artistic Guitarist (see Fig. 5.1).

When presenting The Artistic Guitarist, the guitarist and sound tester was referring to himself, and he shared some of his personal experiences from the music industry. In his explanation of the artistic guitarist he was referring to himself.

He explained, for instance, that he was more interested in the customization and creation of new sounds than being a star. He is not the stage person, but the kind of guitarist who is experimenting with his band in the basement, a person who lives by selling his music to front-line stage-performers via the Internet. By creating and presenting the model shown in Fig. 5.1, the guitarist/sound tester made some of his sticky knowledge about guitarists explicit to the other participants in the workshop. The Artistic Guitarist created a concept of meaning in which he structured all his experiences. Until the workshop, he had never shared any of these insights with anyone at TC Electronic, nor tried to make them explicit. However, the insights about The Artistic Guitarist had an important influence on the workshop and on the other participants' understanding of TC's market.

L. Møller and C. Tollestrup, *Creating Shared Understanding in Product Development Teams*, DOI: 10.1007/978-1-4471-4180-8_5, © Springer-Verlag London 2013

Fig. 5.1 The Artistic Guitarist is shown with his 'yellow stick'-guitar. The guitar is connected to the white guitar pedal as well as to the computer from which different sounds can be created/adjusted or downloaded from the Internet

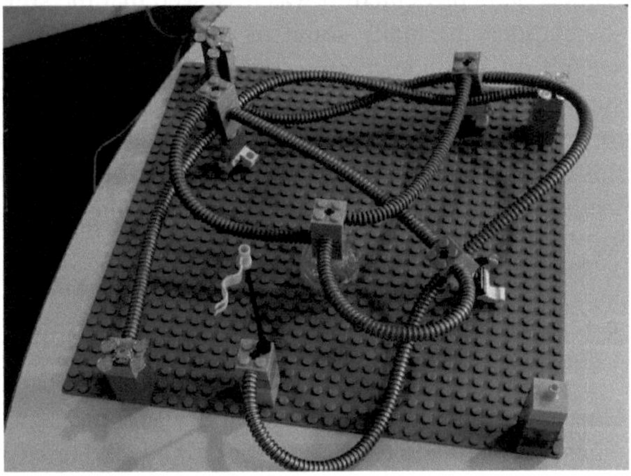

Fig. 5.2 The model called: Running Away From the Task—from workshop 3

Another example that illustrates how the significant models enabled participants to make some of their sticky knowledge explicit is the model called Running Away From the Task, from the workshop at Daimler, shown in Fig. 5.2. It was created by one of the industrial psychologists as a picture of the many different 'behaviour steering mechanisms' like money, department policies and personal interests, which influence engineers working in development teams. The main message was that, in the end, all these mechanisms may ruin team collaboration and make the engineers run in all different directions, instead of focusing on the task and the collaboration.

By presenting this model in the workshop, the industrial psychologist made some of his sticky knowledge about the organization explicit to his colleagues and, as it was reviewed in Chap. 3, it had a significant influence on the workshop and the shared model called 'The Circular Approach'.

In order to understand the reason behind the discovery that the significant models enabled participants to make some of their sticky knowledge explicit, it can be helpful to compare the structure of the significant models with the ladder of inference reviewed in the theoretical framework. The ladder of inference [1] showed that our present beliefs influence the data we select next time—and accordingly, the senses that we allow to influence our personal meaning making are bound by our previous meaning making.

However, in the creation of the significant Lego models, the creator structures all his experiences into a concept or reconnects his senses into a meaning making.

In doing so, the reflexive loop may be stopped, because he is so focused on the process of creating the physical artifact and therefore pays undivided attention to his experiences. This means that when he creates the significant model, he starts on the first step of the ladder instead of on its fifth step, which is probably the step he would have started at in a conversation.

Some of the arguments for this line of events, where the reflexive loop is stopped and the focus is on creating a physical artifact, can be found in Schön's description of the designer's interactions with models and prototypes [2]. When creating the model or prototype, the designer engages in a process of trial and error in which he uses his tacit knowledge-in-action (which is based on previous experiences).

Another argument is that the reflective loop actually seemed to be stopped in the workshop at Red Cross, when they were creating the shared model. As explained earlier, in the process of building their different models of experiences and reflections together, they created The Modular Structure without anyone being fully conscious of it.

In the example of the sound tester/guitarist from TC Electronic, the situation might have been that he did not find it purposeful or appropriate to talk about his passion in his work as a sound tester, and that he had accepted the Jimi Hendrix Guitarist as the main user of TC's products; therefore, he was standing on the fifth step of the ladder of inference, and his actions were based on these beliefs (Fig. 5.3).

Nevertheless, when he created The Artistic Guitarist, he connected all his experiences—including those from outside TC—into this personal model, which means he was starting on the first step of the ladder of inference.

In the example at Daimler, the situation might have been that the industrial psychologist was so used to his colleagues' professional practice in terms of diagnostics, intervention and evaluation, as well as his own that he had not even considered the fact that all this fitted very poorly with his own sticky knowledge about the organization. Still, when he was creating the model 'Running away from the task' to explain the context of the Social System Engineering Program, he was reconnecting and recombining all his experiences from the organization, despite his understanding of the psychological approach.

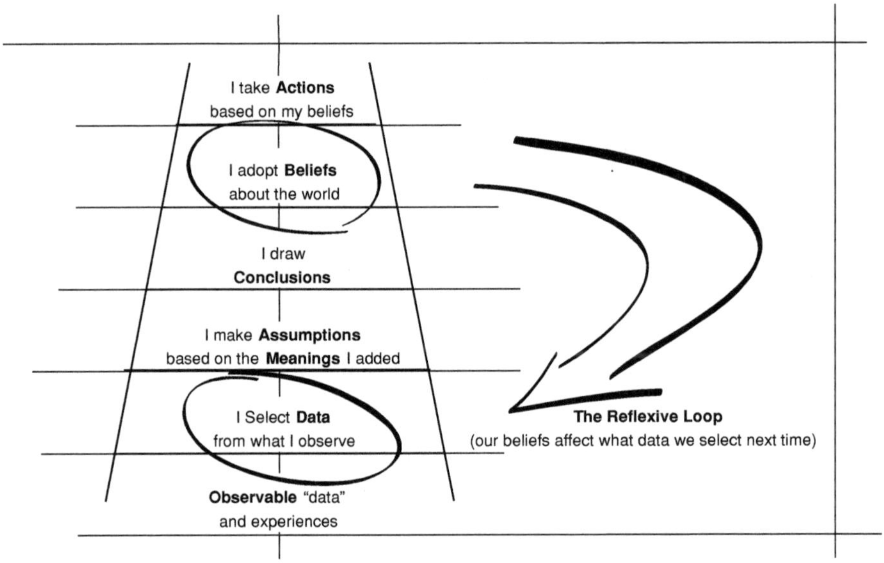

Fig. 5.3 Ladder of inference (Adapted from [3] p. 243)

5.2 Second-Order Understanding of the Creators' Meaning Making

The second finding, which needs to be unfolded, is that the significant models provided a second-order understanding of their creators' meaning making, and thereby supported the communication in the group.

An example of this can be found in workshop 4 and the Medical Treatment House project in the model called: Pulling Together (or directly translated from Danish: 'Lifting in a Crowd') shown in Fig. 5.4.

The model represented the concept of meaning based on the director's experiences: that in order to make a specific Medical Treatment House a reality, it was necessary that the different stakeholders were pulling together. Therefore, the region, the municipality, and a number of practitioners (doctors) would have to pull together their interests and invest together, in order to build the Medical Treatment House. After presenting the model, he further shared some of his experiences from other shared investments between the municipality and the Region Northern Jutland, including insights he had from current cases in Denmark, where Medical Treatment Houses had been created. The presentation of the model 'Pulling Together' revealed to the other participants how the director was making meaning in relation to the project—and thereby, the model made it possible for the other participants to understand his understanding (second-order understanding).

The reason why the significant models provided a second-order understanding of their creators' meaning making and thereby supported the communication in the

Fig. 5.4 The model called: 'Lifting in a Crowd' (Pulling Together) from workshop 4

group can also be revealed by looking more closely at the structure of the significant models. The significant models have a 'concept component' as well as an 'experience component'. This means that they are better at providing an understanding of how their creator understands the situation, because both the meaning and the context for this meaning making are presented. As explained in the previous chapter, when meaning alone is communicated, it can be hard to understand, as there is no context to understand it in; similarly, experiences alone can be hard to connect and link together into something meaningful.

However, the significant models hold both meaning (in terms of a concept) and perceptions and experiences (in terms of the personal storytelling), and thereby they support the communication, because the other participants are able to see or reconstruct the whole meaning-making process (or intellectual moulding) made by the person presenting.

In Fig. 5.5, an illustration is made in relation to communicating meaning and the communication made through the significant Lego models, respectively.

On the left side of the illustration, the participant is only communicating meaning. This meaning derives from previous experiences (senses and perceptions), yet, there is no longer a link between the meaning and senses/perceptions, which it derived from. On the right side, however, meaning as well as experiences are communicated.

5.3 Creating a Shared Frame in Relation to the Project

When analyzing the six workshops, it was also found that a number of shared significant models enabled the participants to create shared frames in relation to the project. In the theoretical framework, a shared project frame was defined as the basis upon which the team matches problems with solutions [2, 4]. The shared frame was described as either a selection of a desired end state or goal, which

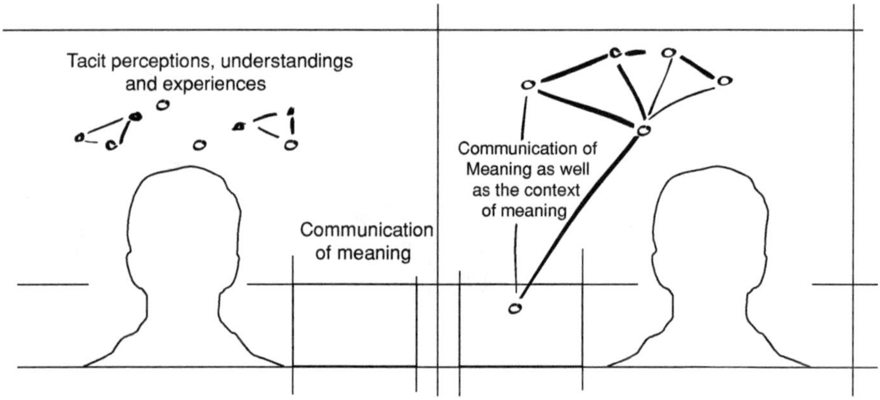

Fig. 5.5 Communication of meaning alone versus the significant model

Fig. 5.6 The shared model from workshop 1

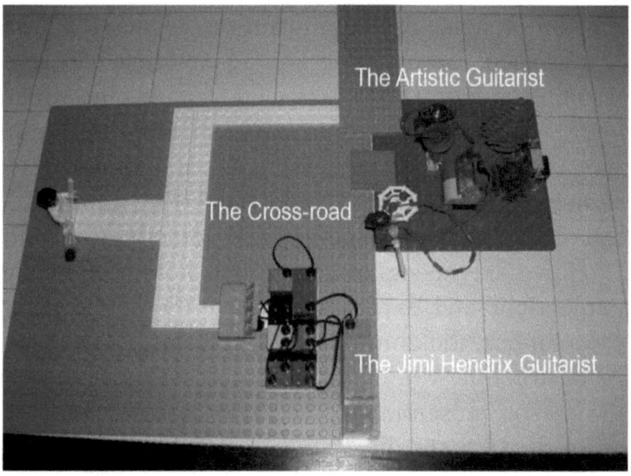

implicitly includes the problem, need, or opportunity. Or an identification of the problem, need or opportunity, which implicitly includes the desired end state or goal. To exemplify that the shared significant models enabled participants to create shared frames in relation to the project, all six workshops will be reviewed.

5.3.1 Workshop 1

In workshop 1 at TC Electronic, the shared model was based on three shared significant models. (1) The Jimi Hendrix Guitarist, (2) The Artistic Guitarist, and (3) The Cross Road, which symbolized the choice between the two types of guitarists.

The shared model (Fig. 5.6) created shared framing of the project in terms of the choice between two different frames along with a shared understanding of these two frames. On the one hand, there was the Artistic Guitarist and his needs in

Fig. 5.7 The shared model with the modular structure from workshop

relation to digital sound (and the opportunity to explore and use the qualities of digital sound technology), and on the other hand, there is the Jimi Hendrix guitarist, who preferred the analog sound.

When the shared model was presented to TC's management after the workshops, it was implied that there was actually only one right choice in terms of the Artistic Guitarist.

5.3.2 *Workshop 2*

In workshop 2 at Red Cross, the shared model was based on one shared significant model called The Modular Structure shown in Fig. 5.7. This divided the base camp into different modules, which could be developed and shipped independently.

This significant model created a shared frame of the project by presenting a desired end state of the base camp, in terms of the modular structure; and it implicitly reveals the understanding of the problems, which had to be solved in the base camp, such as different types of needs, when it comes to heating in the sleeping areas and in the working areas, and the issues in relation to setting up the camp.

5.3.3 *Workshop 3*

In workshop 3 at Daimler, the shared model was a combination of the two significant models called The Linear Approach and Running Away from the Task, which then turned into the model called The Circular Approach (see Fig. 5.8). The Circular Approach represents the Social System Engineering Program and presents the idea

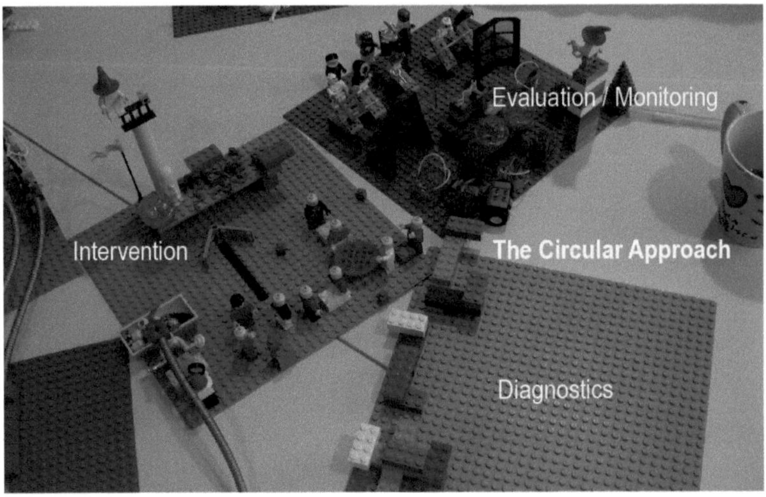

Fig. 5.8 The shared model from workshop 3

to constantly diagnose, make interventions, and evaluate the development teams. The Circular Approach combined both the experiences of the many 'behavioural drivers', which constantly challenge the collaboration in the engineering teams, as well as the approach and experiences linked to industrial psychology.

The Circular Approach created a shared frame of the project in that it shows the desired end state of the Social System Engineering Program. It also implicitly expresses the understanding of the problem in the engineering teams, which has to be approached by this program, in terms of the understandings deriving from the model called Running Away from the Task.

5.3.4 Workshop 4

In workshop 4 and the project Medical Treatment Houses, the shared model was first and foremost represented in the shared significant model called Pulling Together. The model expressed that in order to make a specific Medical Treatment House a reality, it was necessary that the different stakeholders were Pulling Together. In the discussion after the workshop, this model was somehow merged together with the other significant model from the workshop, called The Community Mill. This fusion of the two models extended the idea of Pulling Together to all levels and all times—from a specific Medical Treatment House, where the region, the municipality and a number of practitioners have to pull together their interests, and invest together—to the everyday operation of a Medical Treatment House, in which different people working have to Pull Together to make the place function for the citizens and themselves.

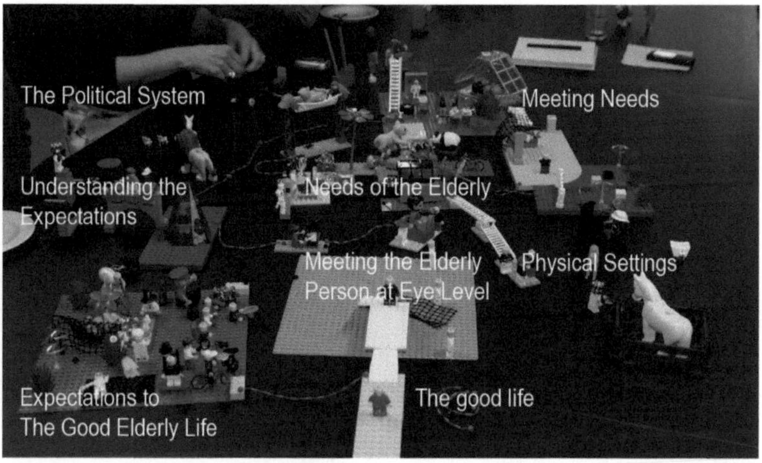

Fig. 5.9 The shared model from workshop 5

The model called Pulling Together created a shared frame of the project in that it represented the desired way to create the Medical Treatment Houses, while it showed implicitly what the largest potential problems were, if the different parties chose not to collaborate and pull the project in different directions.

5.3.5 Workshop 5

In workshop 5, the final model was especially characterized by the four shared significant models: The good Life, Needs of the Elderly, To be Met at Eye Level, and Expectations towards the Elderly Life (see Fig. 5.9).

What characterized these shared significant models was that they all had the elderly person as point of departure and not the nursing home setting. These significant models created a shared frame of the project in that they represented the desired goal of the project, in terms of putting the elderly person in focus. It also highlighted implicit problems in focusing on the nursing home situation only, in that the solutions (product and services), therefore, would be limited by the structure of these.

5.3.6 Workshop 6

In workshop 6, the final model included shared significant models such as: The Prison of Loneliness and Boredom, To Give and to get Help, The Gate to the Elderly life, The Brown Mass, Human Rights, The Crocodile in the Glass Cage, and Structures that can be broken.

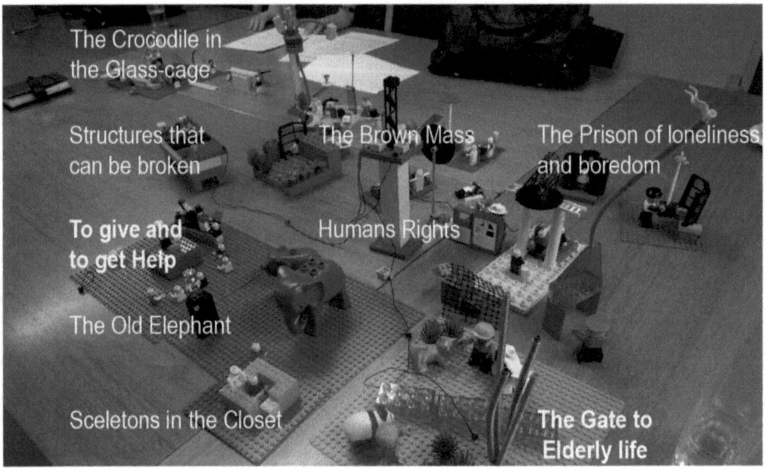

Fig. 5.10 The shared model from workshop 6

These significant models created a shared frame of the project in that they identified some of the key problems to be approached in the project, such as loneliness, getting through the gate, being able to give something back and being in systems, where structures can support one's life, but also be broken. The significant models also implicitly showed what kinds of solutions were needed in the project. After the workshop, it was even suggested that they would be used as design dogmas or design principles (Fig. 5.10).

5.4 The Background

In order to understand, why the significant models enabled participants to create shared frames in relation to the project; it can once again be helpful to investigate their structure. A common feature for all the shared significant models—no matter how they emerge—is that they reflect what happens in a personal meaning-making process: they show how the individual combines and links perceptions and experiences into meaning, due to the 'concept component' and the 'experience component'. In the example of adoption, the other participants chose to invest 'ownership' in this reflection of the participant's personal meaning making, and thereby they initiate the creation of a shared frame. The understanding of how the creator of the significant model is making meaning influences the other participants' personal meaning making, because they are able to see their own experiences in the light of the significant model. Furthermore, they also come to understand and, to some extent, to share the experiences of the creator, whose

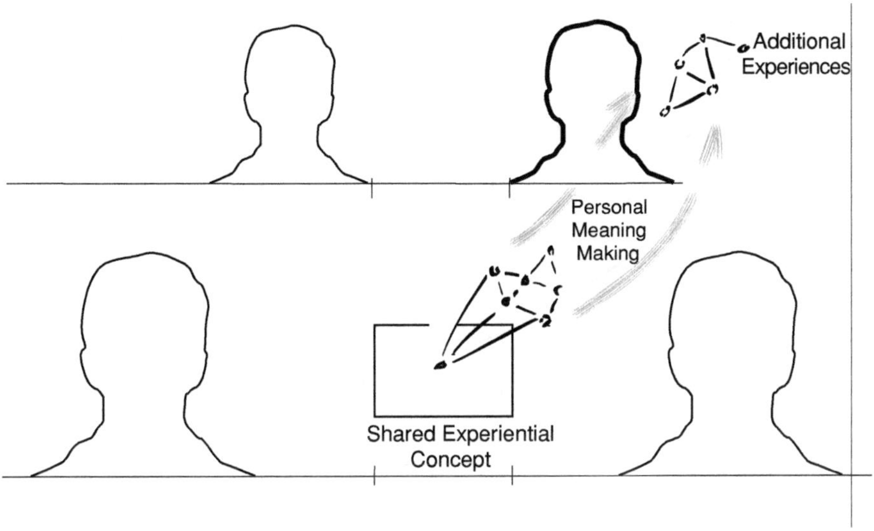

Fig. 5.11 Adopting a shared experiential concept influences the personal meaning making

experiences thereby become part of the other participants' personal meaning-making processes, as well. This can also be illustrated via the example of The Gate to the Elderly Life. As mentioned earlier, the other participants are able to see their own experiences with elderly relatives in the light of 'The Gate' and thereby understand complaints about the food or physical limits as a lack of acceptance and acknowledgment, in relation to being old. In workshop 5, the participants also experienced Eva's personal storytelling. The stories and experiences, which she shared, thus impact on the other participants' personal meaning making (at least to some extent). This is also illustrated in Fig. 5.11.

In the examples where two individual significant models are combined, or where a number of experiences are combined and linked together via a new 'concept component', the participants are playing an active part in managing and reorganizing the different experiences and linking the various senses and perceptions into meaning. This is another way of creating a shared frame within the team, because all the participants are simulating a personal meaning-making process, in that they are combining and linking senses and experiences into meaning. As illustrated in Fig. 5.12 this influences the participants' personal meaning-making process too, because this process equips them with a concept (or a perspective) in which their personal senses, perceptions and experiences can be combined and also with an extended set of experiences (because they now 'share' some of the other participants' experiences, through their personal storytelling).

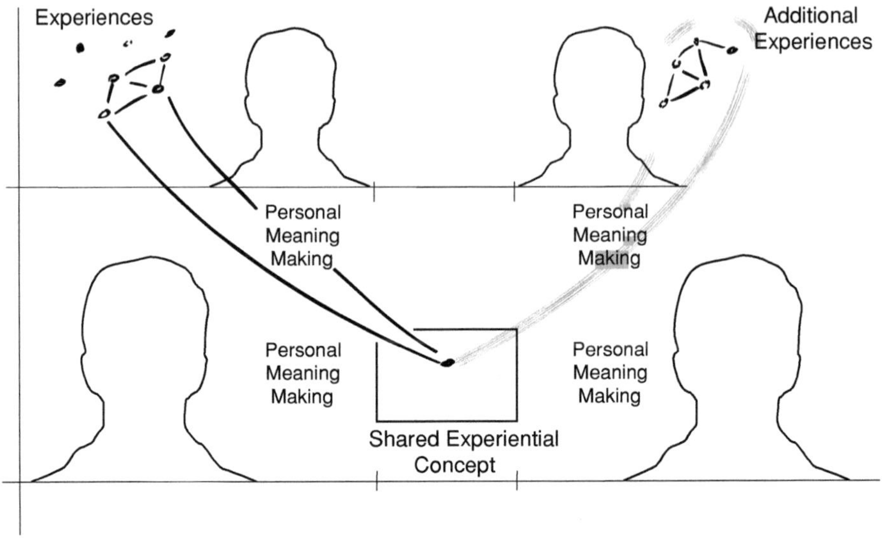

Fig. 5.12 Creating a shared experiential concept and its influence on personal meaning making

5.5 Summary

In this chapter, the general findings in relation to the significant models have been unfolded in terms of examples. It has also been reviewed why the significant models influenced the communication of sticky knowledge and meaning making the way they did, and how they supported the creation of shared frames in the teams.

In relation to this, the structure of the significant models—in terms of a 'concept component' and an 'experience component'—has been identified as very influential.

References

1. Argyris C (1990) Bryt forsvarsrutinene—Hvordan lette organisasjonslæring. University Press, Cambridge (Original title: Overcoming organiza tional defences: facilitating organizational learning)
2. Schön DA (1983) The reflective practitioner. Basic Books, New York
3. Senge P, Ross R, Smith B, Roberts C, Kleiner A (2001) The fifth discipline—fieldbook. Nicholas Brealey Publishing, London
4. Valkenberg R, Dorst K (1998) The reflective practice of design teams. Des Stud 19(3):249–272

Chapter 6
Personal and Shared Experiential Concepts

In the previous chapter the identification and characteristics of the significant models were unfolded and exemplified. This chapter will go a step further in terms of qualifying the significant models definition-wise. A large part of this study has actually evolved around the puzzling question: What are these significant models? Initially, both boundary objects and metaphors seemed to be plausible definitions. However, as it already has been revealed, none of these definitions fully covered the characteristics of the significant models, and it was therefore found necessary to present and identify the significant models as something new and not previously defined. This chapter will start with a review of the comparison between the significant models and boundary objects and metaphors, respectively. This will be done to show why these definitions were disqualified as explanations for the significant models. A naming of the significant models will follow this review. And to underline this, the characteristic elements of the significant models, which have been found in the previous two chapters, will be summarized.

6.1 Exploring Definitions

The first definition, which initially seemed plausible to characterize the significant models, was boundary objects.

6.1.1 Boundary Objects

As reviewed in the theoretical chapter, boundary objects are understood as objects, which enable different groups to see and give meaning to different aspects of the same object, even though they have different backgrounds, competences, practices, or professional languages [3].

L. Møller and C. Tollestrup, *Creating Shared Understanding in Product Development Teams*, DOI: 10.1007/978-1-4471-4180-8_6, © Springer-Verlag London 2013

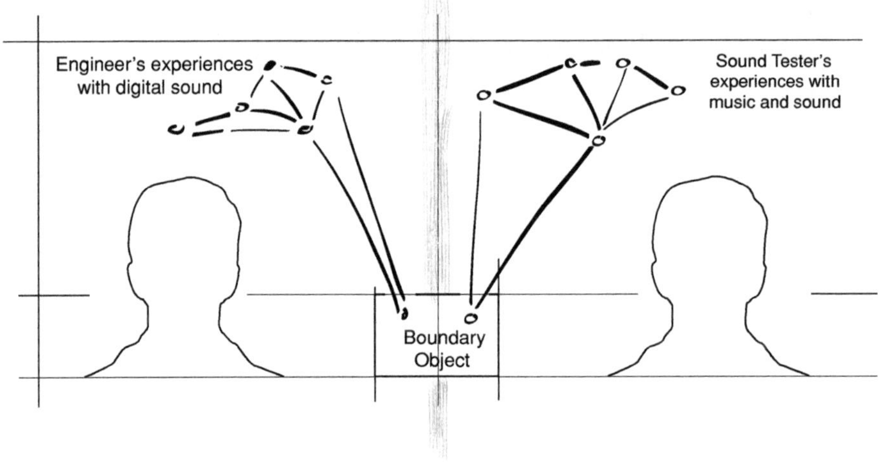

Fig. 6.1 The guitar-pedal as a boundary object

The Boundary Object definition fits very well in terms of creating an object for interaction between people with different backgrounds, competences, practices, and professional knowledge; but Boundary Objects do not necessarily require that participants create a shared framing. As Henderson [1] p. 450) argues:

> (...) Boundary Objects allow members of different groups to read different meanings particular to their needs from the same material. This is possible, because the material remains flexible in group use and more focused in individual site use.

The significant models, on the other hand, were not characterized by allowing different people to read different meanings from the same object; on the contrary, they enabled stakeholders to make their sticky knowledge explicit, provided a second-order understanding of their creators' meaning making, and supported the creation of shared frames.

Workshop 1 actually has a fine example of the difference between the characteristics of the boundary object and of the significant models in the workshop. As described earlier, the part time guitarist worked at TC Electronic as a sound tester, which means that he tested the prototypes of the guitar pedals created by the engineers. In their interaction, the prototypes of the guitar pedals became boundary objects. The guitarist was testing the sound of the prototypes and so was mainly interested in the prototypes from a musician's perspective, whereas the engineers creating the prototypes were mainly interested in the prototypes from a technological perspective. In other words, they read different meanings into the prototypes. This is further illustrated in Fig 6.1.

However, the significant model from the workshop called The Artistic Guitarist was not a boundary object. It was a way for the guitarist/sound tester to communicate some of his sticky knowledge about his work as a musician, and the aim

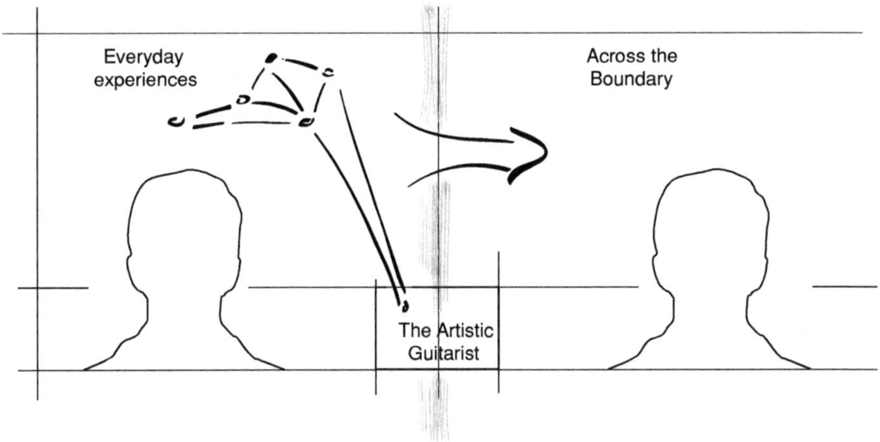

Fig. 6.2 'The Artistic Guitarist'—a way of getting across the boundary

of creating it was to enable the other participants in the workshop to understand his understandings.

In other words, the significant model is not a boundary object, but a way of getting across such boundaries (See Fig. 6.2).

6.1.2 Metaphors

Another attempt to position the significant models into an already existing definition was based on the idea that perhaps the significant models were metaphors. One of the main similarities between metaphors and the significant models is that the significant models often were expressed through metaphors or phrases, like for example The Prison of Loneliness and Boredom. As described in the theoretical framework, Lakoff and Johnson argue that:

> (…) most of our ordinary conceptual system is metaphorical in nature ([2]. p. 4)

This means that metaphors are used to map experiences into cognition and thereby make meaning of them. Initially, this description of metaphors also seemed to fit the characteristics of the significant models.

However, when investigating the video material it was clear that the metaphors only underlined part of the characteristics in the significant models. In other words, the metaphors only represented the 'concept component' of the significant models, whereas the definition of a metaphor does not include an 'experience component'. The fact that the 'concept component' often is metaphoric may be part of the reason why the significant models made the impact on the individuals' meaning making, which they did.

Still, it does not mean that the definition of a metaphor covers the characteristics of the significant models. In the significant model the 'experience component' is what links it to its creator. Therefore, the personal storytelling is an important part of the significant models, because it shows not only their meaning, but also the context based on which this meaning is created.

6.1.3 Naming and Characterizing the Significant Models

After examining boundary objects and metaphors, it is evident that none of these definitions fully cover the characteristics of the significant models. It is therefore found necessary to understand the significant models as something new and not previously defined.

Based on the analysis in the previous chapters, it is however possible to characterize the significant Lego models as well as to give them names that can be used in a wider context.

The name 'Personal Experiential Concepts' and 'Shared Experiential Concepts' emerged after a thorough analysis of the workshop videos and from several discussions in relation to the initial writing process. The names were mainly chosen to underline the 'experience component' and the 'concept component', which were found so particular to these entities. It was further found necessary to show the division between the personal and shared models—and subsequently adding these to the names, as well.

In the following page, the characteristic elements in the Personal Experiential Concepts and Shared Experiential Concepts are summarized.

6.2 Personal Experiential Concept

- A Personal Experiential Concept is a way of communicating a personal meaning making in relation to a project, activity, or context—in terms of how different experiences (senses and perceptions) are combined into personal meaning making.
- A Personal Experiential Concept holds both a 'concept component'—in terms of a concept of meaning (that can be generalized across different people's experiences) and an 'experience component'—in terms of personal experiences.
- A Personal Experiential Concept enables stakeholders to make sticky knowledge explicit, because the unconscious senses and perceptions, as well as the conscious reflections and interpretations (in the form of meaning) are communicated.
- A Personal Experiential Concept is not a defined boundary between different sets of knowledge, perspectives, assumptions, or values, but a way of communicating across such boundaries, due to its two component structure, which supports and enables the receiver to understand the sender's construction of meaning. The Personal Experiential Concept is illustrated in Fig. 6.3.

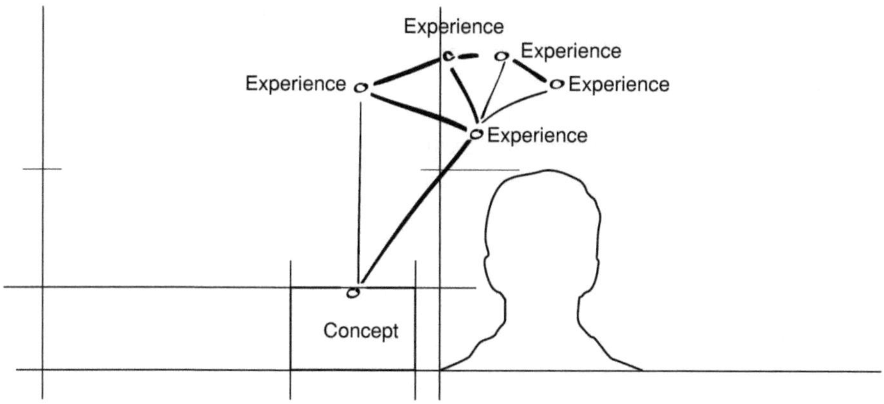

Fig. 6.3 Illustration of a Personal Experiential Concept including 'Concept component' and 'Experience Component'

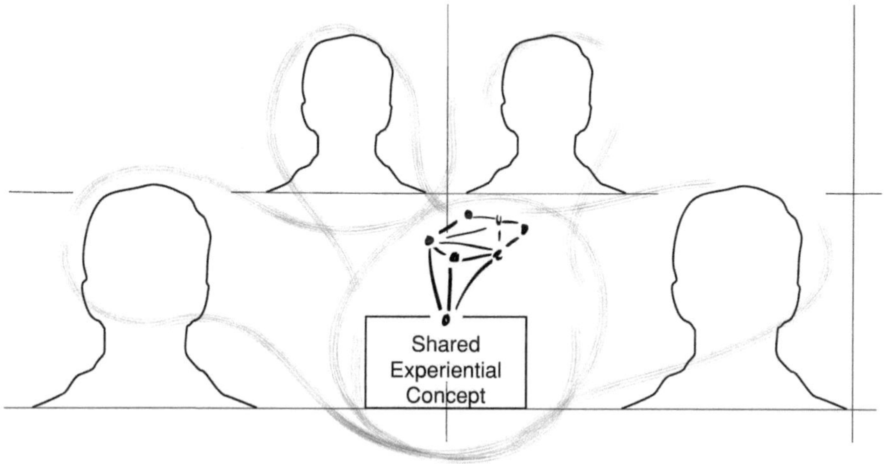

Fig. 6.4 Illustration of a shared experiential concept where all share the concept and experience components

6.3 Shared Experiential Concepts

- A Shared Experiential Concept is constructed by a 'concept component' (which can be generalized across different people's experiences) and an 'experience component', in the form of at least one personal storytelling.
- A Shared Experiential Concept is created, when a design team either adopts a Personal Experiential Concept, or combines two different Personal Experiential Concepts into one, or provides a number of experiences/reflections with a concept, which links them together.

- The creation of a Shared Experiential Concept is a reflection or simulation of a personal meaning-making process, and therefore it influences its owner's personal meaning making.
- A Shared Experiential Concept or a group of Shared Experiential Concepts support and enable the owner's creation of a shared framing. The Shared Experiential Concept is illustrated in Fig. 6.4).

References

1. Henderson K (1991) Flexible sketches and inflexible data bases: visual communication, conscription devices, and boundary objects de sign. Eng Sci Technol Hum Values 16(4):448–473
2. Lakoff G, Johnson M (1980) Metaphors we live. University of Chicago Press, Chicago
3. Star SL, Griesemer JR (1989) Institutional ecology, 'translations' and boundary objects: amateurs and professionals in Berkeley's Museum of Vertebrate Zoology, 1907–1939. Soc Stud Sci 19:387–420

Chapter 7
Conclusion

This chapter holds the conclusion of the study. The aim is to synthesize the different theoretical and empirical parts of the book, and to connect the different perspectives and insights shared. The aim of this chapter is more specifically: To finish the research circle by answering the research questions presented in the introduction and outline its limitations. To reunite the theoretical foundation with the outcome of this research project, and thereby position the present conclusions within existing knowledge. And finally, to discuss the perspectives and further implications from this study. This chapter is divided into smaller sections, in which these different aims will be discussed and wrapped up.

7.1 Answering the Research Questions

The answers to the research questions have, to some extent, already been reviewed in the empirical chapters. However, in this chapter the questions will be directly connected with the answers.

This study has evolved around a very specific research setting—in terms of the facilitated workshops, in which the creation of physical artifacts is a central element. As it has been reviewed, all the workshops took place in projects positioned in the early phase of innovation and included both members of the interdisciplinary team, as well as users and stakeholders with the relevant contextual knowledge. The aim of the workshops was to create a shared framing of the project or to find the right problem, need, or opportunity to approach.

In this book, the research questions are also framed within this specific workshop setting, and their answers are to be understood in this context as well.

The Research Questions are:

1. How can the creation of physical artifacts enable and stimulate the communication among team members, users and stakeholders in interdisciplinary teams working in the early phases of innovation?

L. Møller and C. Tollestrup, *Creating Shared Understanding in Product Development Teams*, DOI: 10.1007/978-1-4471-4180-8_7, © Springer-Verlag London 2013

2. How can the creation of physical artifacts enable and support the creation of shared frames within interdisciplinary teams working in the early phases of innovation?

7.1.1 Research Question No. 1

Based on the research presented in this book it is found that:

One of the ways in which physical artifacts can enable and stimulate the communication between team members, users and stakeholders in interdisciplinary teams working in the early phases of innovation, is in the form of Personal Experiential Concepts.

The main reason, for this, is that Personal Experiential Concepts enable team members, users and stakeholders to communicate how they make meaning in relation to a project or their everyday life, and thereby enable them to make some of their sticky knowledge explicit.

Secondly, Personal Experiential Concepts enable the interdisciplinary team to create second-order understandings of each other's perspectives, values, and assumptions. These second-order understandings illuminate any differences in the participants' understandings in relation to the project, and make it possible for the participants to discuss or negotiate these.

The reasons why Personal Experiential Concepts facilitate these things are their two-component structure. As reviewed in Chap. 6, a Personal Experiential Concept holds both a 'concept component' and an 'experience component'. The 'concept component' contains a concept of meaning, which can be generalized across different people's experiences. The 'experience component' holds a number of personal experiences, which are shared in the form of storytelling.

Together, these two components become a way to communicate personal meaning making in relation to a project, activity or context, because they show how different experiences (senses and perceptions) are combined into personal meaning making. Or in other words, they are important drivers for the communication of personal meaning making, because they reveal both meaning and the context of this meaning.

7.1.2 Research Question No. 2

Based on the research presented in this book it is found that:

One of the ways in which physical artifacts can enable and support the creation of shared frames within interdisciplinary teams working in the early phases of innovation, is in the form of Shared Experiential Concepts.

In the early phases of innovation, Shared Experiential Concepts enable teams to create a shared frame in relation to the project, because the creation of a Shared

Experiential Concept resembles or simulates a personal meaning-making process, and thereby influences the team members' individual meaning making, by providing a model of how to structure different people's experiences into meaning.

Like in a Personal Experiential Concept, the two characteristic elements of a Shared Experiential Concept are the 'concept component' and the 'experience component'. However, in a Shared Experiential Concept both components are shared among a group of people within a team. This means that the 'concept component' is shared in terms of a shared concept of meaning, and the 'experience component' is shared through personal storytelling.

The team either (1) adopts a Personal Experiential Concept, (2) combines two different Personal Experiential Concepts into one or (3) provides a number of experiences/reflections with a concept, which links them together—in order to make them shared.

Furthermore, the creation of a Shared Experiential Concept is a shared process of experiencing in itself, because the team members are sharing the experiences in the workshop and the experience of building their Personal Experiential Concepts together into one.

By creating a Shared Experiential Concept and thereby a shared frame, 'the sharedness' in the team, which is highly recommended in previous research, is also created. This limits the possibility that the team ends up pursuing different goals, as well.

7.2 Limitations of the Conclusions

Based on the research material and research methods used in this project there are a few important limitations of these conclusions to note.

7.2.1 Limitations of the Research Approach

One of the main limits to the research approach is the fact that most of the workshops were conducted with the researcher as the facilitator, and that the workshops were presented as part of a research project (hence the video camera).

The mere presence of the researcher may have created a feeling of 'being observed' and thereby changed the participants' actions and reactions toward each other. In favor of the study, it can be argued that in the introduction to every workshop an effort was made to underline the fact that the research did not involve any judgment of individual behavior, and that the attention was on the models. However, the researcher's potential influence on the participants' behavior cannot be ignored.

In relation to the discussion on the role of the researcher, it has to be noted that the identification of Personal and Shared Experiential Concepts did not appear until after the first five workshops were conducted (that is all workshops except workshop 4). This implies that the researcher could not intentionally have pushed the development of Personal and Shared Experiential Concepts in regard to the largest part of the research material.

7.2.2 Limitations of the Documentation Approach

Another thing which has to be discussed in relation to the reliability of the research project is the presence of the video camera. Although people adjust to a video camera rather quickly [2], its presence may have had an effect on the participants' actions and reactions and thereby limiting the realism of the data. When looking at the data from the workshops, it is evident that people focus on the interactions in the group, and not on the camera. However, there is a possibility that they may have been holding themselves back or somehow limited their behavior, due to the presence of the camera.

7.2.3 Limitation by Empirical Data

The empirical material only represents one longitudinal study that provided the opportunity to follow the participants and the development of a project. Without follow-up interviews on the workshop, there is not enough data to support the long time commitment and continuous interpretation of the Personal and Shared Experiential Concepts developed and defined at the workshops. Hence, the conclusion that Personal and Shared Experiential Concepts support the team- and stakeholder communication and the creation of shared frames in the early phases of innovation is not well supported outside the workshop setting, in which they are created. However, strong indications from the longitudinal study point towards these Personal and Shared Experiential Concepts playing a role beyond the workshop stage (See Sect. 7.4).

7.3 Reuniting the Theoretical Foundation With the Findings

Throughout this book, an effort has been made to connect the design perspective with the development of the workshop setup as well as to connect the insights from the theoretical framework with the empirical findings. However, the connection between the outcome of this research and the previous research in the early phases of innovation has not been unfolded yet.

In the introduction, it was clarified that this research builds on the a priori assumption that all interdisciplinary project teams working in the early phases of

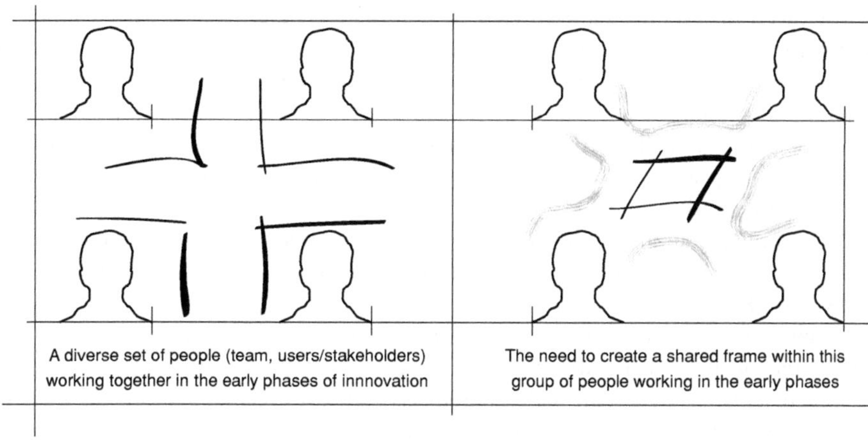

Fig. 7.1 Positioning the findings

innovation will benefit from a shared frame as early as possible, even though this framing may be changed to another shared frame later in the process.

This assumption is linked to previous research in teams, which shows that 'sharing' is one of the keywords, when it comes to handling diversity, complexity, and ambiguity in teams. But in the research setup in this project, it is also linked to the research on how to handle the asymmetry or stickiness of knowledge in the early phases of innovation, in the sense that if the interdisciplinary team works with users and stakeholders to create a shared framing of the project, it is assumed that this project framing includes the most important problem, need, or opportunity from the context of use.

This research can, therefore, be seen as an effort to approach the gap between a diverse set of people (team, users and stakeholders) working together in the early phases of innovation, and the need for these people to create a shared frame. This is also illustrated in Fig. 7.1.

This study is important because it reveals some of the underlying drivers for enabling and stimulating communication within this group of people, as well as some of the drivers for enabling and supporting the creation of a shared frame. In other words: Personal and Shared Experiential Concepts are the link between 'the diverse set of people' and 'the need to create a shared frame' or an answer to the question: How can these people create a shared frame together?

Accordingly, the main contribution of this research is the identification of Personal and Shared Experiential Concepts and their influence on meaning making and on the creation of shared frames in interdisciplinary teams working in the early phases of innovation. The notion of Personal and Shared Experiential Concepts and their structure has not been unfolded previously, and it is therefore considered the 'breakthrough' of this book.

On a broader scale, the aim of this research project was to approach the gap between the internal and the external challenges in the early phases of innovation.

As reviewed in the introduction chapter, previous research on the early phases of innovation either focuses on how to handle the diversity, complexity or ambiguity within the team, or on the interaction between users and stakeholders and how to manage the asymmetry or stickiness of knowledge.

It was found that this division did not exist in practice, and that both sets of challenges are present at all times in interdisciplinary teams working in the early phases of innovation.

Accordingly, the aim of this research became to approach the gap in the present knowledge in terms of the intersection between the internal- and external challenges (see Fig. 1.3).

When evaluating the outcome of this research, it is clear that it brings forward knowledge on how to handle both the internal- and the external challenges. As explained above, Personal and Shared Experiential Concepts are a way to approach the diversity, complexity, ambiguity and stickiness of knowledge, which the teams have to handle.

However, this research represents only one way of approaching this intersection between the internal- and external challenges, as well as only one way of approaching the situation where a group of people needs to create a shared frame. In fact, this research project has opened up a number of research potentials and new questions. These will be reviewed and discussed in the next and final section.

7.4 Perspectives and Further Research

Like many other research projects, this study has opened more questions, than it has been able to answer, and provided more opportunities to unfold and investigate new ideas, than it has closed, and narrowed down. First of all, the view on the early phases of innovation as a double challenge opens up a number of questions on how to handle different perspectives and methods, which may be applied to approach either diversity, complexity, and ambiguity, or the asymmetry/stickiness of knowledge.

Second, this research project can be seen as a first step within the field of Personal and Shared Experiential Concepts. This means that a number of perspectives and questions can be further unfolded and also that indications from this study need to be further researched.

And finally, a number of questions in relation to the role as a facilitator have become increasingly apparent throughout this research project.

7.4.1 Perspectives on the Role of the Facilitator

One of the first questions occurring in relation to facilitating the workshops is: How do you find the 'right' or most relevant users and stakeholders? This question is crucial because without key users and stakeholders, you are not able to get the relevant information or insights in relation to the project. In the workshops from this research project this question has been solved quite pragmatically, either by asking everyone in the interdisciplinary team to decide who they consider relevant, or by thinking through the context of use, and trying to figure out all the important stakeholders. But in the end it was always defined by who was available.

The second question that faces the facilitator is: How do you motivate participants to share their insights, if it means that they have to share quite personal stories? Or in other words: How do you create a 'safe place' in which vulnerability can be handled? This question became especially apparent after the workshops in the Good Elderly Life project, because it was obvious that the insights shared were very important to the project, but at the same time also highly personal to the elderly people sharing them. Another question also related to this is: How do you manage the balance between the intentions of the workshop and the 'slowness' embedded in letting the participants get to know and trust each other?

A third set of questions, which came up during the process of facilitating the workshop, was actually a result of the democratic structure of the workshop. In more than one case it happened that the power structure in the team collapsed. Not many of the participants noticed it, but the project managers did. How do you prepare the managers for this? And: Is it possible to control if and when it happens?

The final set of questions opened up in relation to the workshops is: How do you document the workshops? And: How do you transcribe the values of openness and democracy from the workshops into its documentation? In this research project, this was only tested in some of the workshops—in terms of creating powerpoint documents with pictures and transcripts from the workshop. However, the interactive intentions with this documentation were never unfolded.

7.4.2 Perspectives on the Personal and Shared Experiential Concepts

The second group of questions, which emerged during this research project, was in relation to Personal and Shared Experiential Concepts. The first and perhaps most interesting questions is: To which extent are the bricks needed? And to which extent is the physical artifact needed?

Based on the experiences from this research project, the best guess is that other physical artifacts could be used instead of the bricks, perhaps wooden toys or clay; these would, however, demand better skills building competences in the beginning of the workshop, as well as some kind of intervention in relation to how the different models are put together.

In relation to the question whether the physical artifacts are needed, the experiences from this project indicated that they are needed, partly because they assure that perspectives and insights are not forgotten, even if the discussion shifts to something else (since the physical artifacts remain on the table). And partly because the knowledge-in-action, which was often applied when creating or building the models together, may not be triggered in a discussion.

Another set of questions with respect to Personal and Shared Experiential Concepts is related to whether they created more than a shared framing within the teams? Did they also create trust and commitment?

The main indication behind this question emerged in The Good Elderly Life project. In the evaluation meeting after the two workshops, head of Copenhagen Living Lab, Thomas Hammer-Jakobsen, described the positive effects of the workshops by stating that, in contrast, to many of their other early phase projects, this project had not had any 'refluxes', where the team members, due to different understandings of the projects, would start pursuing different goals; this meant that the process at some point had to be stopped and started all over again.

'The fact that the project did not have all these 'refluxes' created calmness, security and focus on the project, which again made the following process (after the workshops) more effective. There are lots of things in this project that we do not know, but we agree on the focus of the project'. [1]

He further argued that the workshops had not only created a shared frame, but also a commitment to the shared frame, and that the team in the subsequent process seemed to understand each other's actions. Based on the data from this research project, it was not possible to conclude whether this experience was present in all the workshops or only in this particular case, since The Good Elderly Life project was the only longitudinal study. It was also not possible to identify, why the workshop turned out to have that effect. One guess is that the creation of the Shared Experiential Contexts—in terms of the physical artifacts—may have touched a greater variety of intelligences (as described by Howard Gardner) than a typical conversation, and thereby made it possible for the different team members to connect and commit to each other. However, this is not observable on the basis of this data collection, and therefore needs further research.

Another set of questions in relation to Personal and Shared Experiential Concepts is linked with the conditions for their creation. That is: Are certain conditions in the workshops key to the creation of Personal and Shared Experiential Concepts? And can the natural occurrence of Personal and Shared Experiential Concepts be boosted?

Based on the experiences from the workshops the assumption is: Yes. However, it is trickier to answer how this can be done.

The main experience behind this assumption is actually a workshop, which was not selected as part of the research material. The reason for deselecting it was that this workshop did not have any Personal and Shared Experiential Concepts, and it was therefore found very difficult and confusing to communicate as part of this study.

It also did not conclude with a shared frame, which means that it did not contradict any of the conclusions of the study.

However, in relation to the questions above, it provides an initial opportunity to compare the interactions and conditions for creating Personal and Shared Experiential Concepts with the interactions and conditions, when that is not happening.

One of the things, which seemed to differentiate the 'deselected' workshop from the other workshops, was a tendency in the workshop to leave the concrete matter (the bricks) on the table in favor of a more typical meeting-style discussion. This also induced longer dialogs and debates, with fewer participants involved in the discussion.

In the workshops presented in the book, there was a more rapid dialog, which included all the participants at all times. Sometimes, it was even as if the critical decisions were formalized by a consensus assurance in the group, and subsequently the shared model was not altered, until all participants around the table had given their acceptance of the change.

However, the indication presented above is far from being sufficiently conclusive or elaborate, but just an indication.

Yet, the understanding of the interactions and conditions for creating Personal and Shared Experiential Concepts is definitely interesting in relation to future research.

7.4.3 Perspectives in General

As it has been reviewed in this chapter, the creation of Personal and Shared Experiential Concepts opens a number of questions and perspectives in relation to the early phases of innovation. It touches upon several specific issues regarding team dynamics in general, while pointing towards new potentials for improving interactions in interdisciplinary teams.

It approaches some of the key challenges in relation to interaction with users and stakeholders, and indicates new ways of collaborating and co-developing.

Furthermore, the ability of Personal and Shared Experiential Concepts to improve communication and establish shared frames also makes it relevant to areas, where either framing, meaning making, or communication in groups occur—in general.

In other words, there are possibilities for refinement within the present methods and definition, as well as potential in relation to developing the methods and specifications for other contexts of use, i.e. in other project forms or in other group formations.

In fact, Personal and Shared Experiential Concepts can be seen as a potential point of departure, wherever it is important to align and create a collaborative platform for a group of people—and to establish a common objective and shared understanding.

References

1. Copenhagen Living Lab (2008) Meeting minutes. In: Evaluation meeting I, 24 April 2008
2. Jordan B, Henderson A (1995) Interaction analysis: foundations and practices. J learn sci 4(1):39–103

Chapter 8
Facilitation Guide

This chapter holds a facilitation guide, which shows how to plan and execute a team or stakeholder workshop in the early phases of innovation (in terms of the team and stakeholder application). First, the chapter provides an overview of the project types where a team or stakeholder workshop is relevant, and a number of hands-on steps for planning, executing, and documenting a workshop. Second, the chapter provides guidelines for the facilitator's role and mindset in a workshop, including instructions, facilitations, and interventions, and it unfolds some of the critical interactions and situations, which may occur. And finally, the chapter shows how the facilitator can encourage the development of Personal and Shared Experiential Concepts and use these as main results from a team or stakeholder workshop.

8.1 Project Types and Positioning

When you are working in the early phases of innovation, you are often confronted by abstract and strategic questions. In the effort to answer these, you may find yourself in a process characterized by constant shifts between exploring user needs, potential markets, and new technologies, and framing these insights into an increasingly more precise understanding of what to do next.

The many unanswered questions and shifting processes can make it very difficult to plan and even foresee the next step in the early phases of innovation, and typically, plans are constantly adjusted or even created ad hoc.

In order to make it easier to introduce a team or stakeholder workshop in the early phases of innovation and to help project managers to position such a workshop, the section below shows four different situations (Fig. 8.1), based on the empirical findings, where introducing a team or stakeholder workshop is valuable.

L. Møller and C. Tollestrup, *Creating Shared Understanding in Product Development Teams*, DOI: 10.1007/978-1-4471-4180-8_8, © Springer-Verlag London 2013

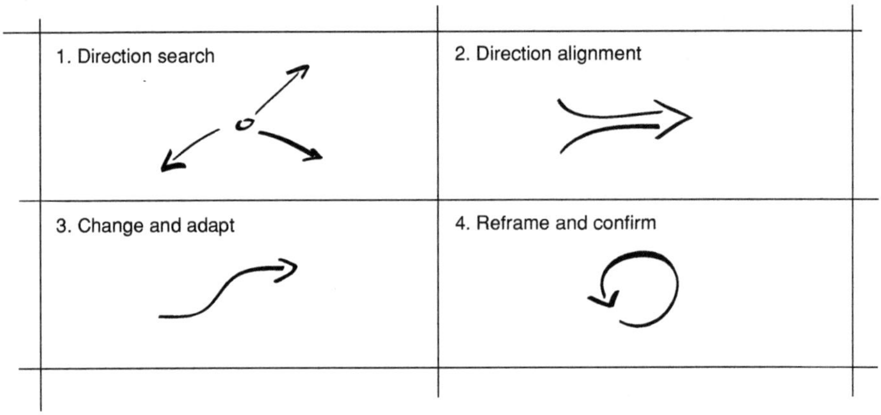

Fig. 8.1 The four relevant situations

8.1.1 Situation 1: Direction Search

When an organization is facing the challenges of new business opportunities or looking for new strategic initiatives it is a situation of “Direction search”.

The situation is best described as pre-project or presearch. It is a situation where strategic and future-oriented questions keep reappearing in the organization, at some point sparking new initiatives to understand in which direction the company should develop.

Workshop 1 (TC Electronic/Guitar pedal) is an example of this type of situation. At the time of the workshop, there was no project setup, nor allocated team or resources; there were some questions within TC Electronic concerning the future of their products, which had encouraged TC’s management to participate in a strategy process together with a nearby university ‘lab’. Part of the lab’s approach to the strategic questions was a stakeholder workshop.

In other words “Direction Search” calls for a stakeholder workshop, in order to get a shared understanding within an organization of the problems, opportunities and needs present in the users’ world, before any new initiatives or R&D projects are setup.

8.1.2 Situation 2: Direction Alignment

When an organization sets out in a new direction or begins a new project with either new or existing members of that organization, it is important to be precise and clear about the goals. This situation calls for “Direction alignment”, which is especially important in the beginning of many early phase projects. At this point in time, the project is only briefly defined, and it is important that the team does not start the project by working in different directions.

Workshop 5 (The Good Elderly Life I) is an example of a direction alignment situation. At the time of the workshop, the project had been described in terms of a funding application, and the Danish Enterprise and Construction Authority had decided to support the project. However, the project was still very open in terms of the scope. Due to the project manager's previous experiences with project teams working in different directions (and thereby loosing valuable time and resources), it was decided to introduce a team workshop.

Workshop 3 (Daimler AG) is another example of a direction alignment situation. The only difference from workshop 5 was that to some extent, the team had been working with the project in different directions at the time of the workshop. Therefore, the team workshop became a needed break in this process and the creation of a shared understanding of the project in the team.

In other words: The 'direction alignment' situation calls for a team workshop in the very beginning of early phase projects, especially if the project holds team members, who have not worked together before, or if the project is cross-organizational.

8.1.3 Situation 3: Change and Adapt

When an organization such as a project team faces new challenges, encounters ground breaking information, or spots new opportunities during a project, it needs to "Change and adapt". This situation can be identified by (1) an emergent user need, (2) a specific problem, or (3) a business opportunity, which arises within a market or within an organization.

Furthermore, there is a demand to approach the need, solve the problem, and take advantage of the opportunity fast, even though doing so can be both ambiguous and complex.

Workshop 2 (Red Cross/Base Camp) and Workshop 4 (Region Northern Jutland/Medical Treatment Houses) are examples of this kind of situation.

In Red Cross, the assignment was to develop and provide a base camp for aid workers in the event of a disaster, and to have it ready as fast as possible. The disaster manager was asked to respond to a need inside the organization, which had great influence on Red Cross's ability to manage disasters. Likewise, the Region of Northern Jutland was asked to solve the problem of a lack of doctors in the region.

Even though dealing with quite different problems, both Red Cross and the Region Northern Jutland responded to the demand by setting up a project team and started solving the problem. And a workshop with stakeholders was planned and executed as part of this.

In other words: The 'change and adapt' situation calls for a stakeholder workshop in order to make the problem in question less ambiguous and complex—and to identify the underlying needs of the different users and stakeholders.

8.1.4 Situation 4: Reframe and Confirm

When an organization has worked 2–3 months in an "early phase project" and gained more knowledge about the project theme, it may need to "Reframe and Confirm" the situation. At this point, the interdisciplinary team has been working with the project long enough to understand the ambiguity and complexity of the project. But the team members are still searching for the underlying needs or 'right problems' to approach.

Workshop 6 (The Good elderly Life II) is a good example of a 'Reframe and Confirm' situation. At the time of the workshop, The Good Elderly Life project had matured and the team had initiated its collaboration; however, a shared understanding of the underlying needs and the 'right' problem to approach in the nursing home context, in order to enhance the life quality of the elderly, was still missing.

In other words: The 'Reframe and Confirm' situation calls for a stakeholder workshop with the objective to identify the underlying needs of the different users and stakeholders. Furthermore, the workshop should create a shared understanding among the stakeholders on what the 'real' problem is.

8.1.5 Other Situations

The four descriptions above are all examples of situations, in which a team or stakeholder workshops were successfully applied. These may be valuable and relevant in other situations, as well. It all depends on the setup of the project, and the adjustments of the workshop to fit the specific situation.

However, the following description of the workshop planning process and the facilitator's role and interventions will focus on the four situations described above and empirically tested.

8.2 Planning the Process

Miguel De Cervantes once said: '*To be prepared is half the victory*'. This is also true regarding a team or stakeholder workshop. Planning the workshop is a very important part of the process and the key to its successful outcome. There are many pitfalls and potential conflicts to address before the workshop, like things, which may not seem important beforehand, but which can end up disturbing the focus of the workshop, simply because the different participants' expectations of the workshop do not fit with its aim and potential. This can be avoided by being well prepared.

8.2.1 Getting the Full Accept From the Manager

One of the first things to do, after having identified a situation suitable for a team or stakeholder workshop, is to gain the full acceptance of the workshop from the manager in charge (whether it be a project manager, a department manager, or the company manager).

There are two reasons for this. First of all, the democratic basis of a team or stakeholder workshop has a tendency to break down the usual power structures within a project team (or within a group of colleagues) during the workshop. For the manager this can be very uncomfortable, especially if he/she does not expect it to happen.

Second, it is important to have the manager's full acceptance, so that he/she can incorporate the result of the workshop into the project.

This means that getting the project manager's full acceptance of the workshop includes his acceptance of suspending *management* of the project during the workshop, taking part in the democratic process, and making sure that the results will be integrated into the project after the workshop, when he/she is back in charge.

Without this acceptance, the workshop may not create the intended value and can create tensions between the workshop participants, in the worst case.

8.2.2 Practical Setup

Having gained the full approval from the manager/project manager, the time has come for the practical planning of the workshop. As in any other facilitated workshop, keeping the participants' focus on the project is the key. The easiest way to achieve this is by planning a whole day workshop, in which the participants are asked to participate from beginning to end. Another way to help the participants to focus on the workshop is by making sure that the needs for breaks, food, and beverages are fulfilled.

In order to support the building together part of the workshop and ensure that the participants hear each other's points of view, dividing participants into smaller groups is not recommended. Preferably, the workshop is held in a room with one large 'conference table', around which all the participants can sit, and with a few extra tables, where the LEGO bricks can be placed. It saves time, if the bricks can be approached from both sides of the extra tables, when the participants are picking up the bricks.

8.2.3 Workshop Participants

Parallel to the practical setup, the selection of the workshop participants takes place. Ideally, a workshop has between 7 and 12 participants. If there are less than seven participants, there is a tendency to have a less elaborate unfolding of the workshop topic, because of the low number of perspectives. And if there are more than 12 participants, it makes the process of building and reviewing each model to the group very time consuming. More than 12 participants also make the workshop seem very protracted, and it can be hard for the facilitator to keep all the participants engaged.

The specific selection of the participants depends on the type of workshop, the situation, and the project, of course. Typically, a Team workshop holds the project team members and the project manager. A stakeholder workshop typically holds the project team as well as users and stakeholders. The users and stakeholders are selected in such a way that they represent as many relevant perspectives as possible, but with extra focus on participants with end-user experience. Especially, in the area of technology development it is further recommended to have both regular users and early adapters.

If the project team and the relevant users and stakeholders exceed the number of 12 participants, it is still recommended to have all the team-members participate in workshop, since it provides valuable insights for them to share. One way to manage this is to ask the team before the workshop to become 'active observers'—by making the explanations of their own models very short, and focusing their attention on the users' and stakeholders' perspectives, instead.

Having actual observers at a workshop is not recommended. It has a tendency to make the stakeholders hold back some of their more personal or tabooed insights. For more inspiration and examples on stakeholder selection please see the review of the different workshops in Chap. 3.

8.2.4 Workshop Questions

Parallel to the practical setup and the selection of the participants, it is time to develop the questions for the workshop. In general, team or stakeholder workshop questions are: Open ended and explorative (cf. wicked problems) and asking for the participants' experiences/understandings, rather than attitudes.

In the development of the questions, the facilitator does not take any explanations of problems or phenomena for granted, as questioning these problems or phenomena can very well be part of the main task at the workshop.

As reviewed in Chap. 3, the nature of the questions also depends on the type of workshop and its objectives.

8.2.4.1 Questions for the Team Application

The objective of the questions in the team application is to ascertain that participants have the same understanding of the project. This means that most project names or themes are used directly in the workshop questions, for instance:

What is the X-project about? What kind of issues and challenges is the X-project approaching?

Besides providing a shared understanding of the project, the objective of the team workshop is also to get a shared understanding of the context of the project. This can be done by asking this question:

Who are the main actors, which influence the X-project?

In workshop 5 (The Good Elderly Life I) the workshop questions were:

What is the Good Elderly Life?

Who are the main actors, which influence the Good Elderly Life?

The questions in the Good Elderly Life took advantage of the ambiguity in *'the good elderly life'* as a phenomenon, which both referred to the project (project name) and to an elderly person's life quality, in general.

8.2.4.2 Questions for the Stakeholder Application

The objective of the questions in the stakeholder workshop is creating a shared understanding of the 'right' problem, challenge, or opportunity to approach. Therefore, questions relate to the users' and stakeholders' understandings and meaning making in relation to their everyday life. The questions are open, but still focused enough to search for the participants' underlying and tacit knowledge. The questions are most often directed at activities, problems, and challenges in relation to the users' and stakeholders' lives, for instance:

What are the main activities in X's daily life? What kind of problems does X experience in daily life? What are the main challenges of X in a Y situation?

Furthermore, it can be valuable to extend some of the questions into a future or ideal situation, for instance by asking:

What would X's daily life look like in 2017? What would the ideal daily life for X be like?

This often brings forward some of the opportunities which do not come about, when focusing on the present situation.

In workshop 4 (Region Northern Jutland/Medical Treatment House) the questions were:

What are the typical problems a patient/citizen may experience in relation to the present healthcare offers?

What should a medical treatment center look or be like to be of interest to a professional practitioner?

When the questions are made, presenting them to the manager in charge can be a good idea, to make sure that the questions match the manager's expectations of the outcome of the workshop.

8.2.5 Facilitator's Role and Mindset

Another thing for the facilitator to prepare before the first team or stakeholder workshop is his or her own mindset. To some extent, the facilitator 'adopts' the non-judgmental, freethinking, and democratic values and represents these during the workshop.

In general, the facilitator's role in the workshop can be described as neutral. He or she focuses on the process of the workshop, the interaction between the participants and the coordination of activities, rather than focusing on the content. This means that the facilitator does not bring any personal opinions into play, but strives to understand and unfold the participants' understandings.

The democratic basis of team or stakeholder workshops also means that the facilitator sees every participant and every perspective as equally important. He encourages participants to share personal understandings and experiences, which can promote the other participants' second-order understanding.

Besides this, the facilitator's role is to guide, motivate, and encourage the participants. Some participants need to be reassured that they can handle the task or that their insights are relevant to the question. Others may need the assignment to be repeated a few times, while others again may need help with specific brick combinations.

Especially when facilitating the first team or stakeholder workshop, it is recommended that the facilitator has participated in a team or stakeholder workshop beforehand and in that way has experienced its qualities or, alternatively, has an experienced facilitator help plan and Co-facilitate the workshop.

8.3 Workshop Instructions and Facilitation

When all the planning activities have been accomplished, it is time for the workshop. As reviewed in Chap. 3, a typical team or stakeholder workshop is divided into four parts: skills building, individual modeling, building together, and reflection. Together these four parts provide the number of steps the facilitator guides the participants through to complete the workshop.

8.3.1 Part 1: Skills Building

The main purpose of the skills building part is for the facilitator to set the stage and prepare the participants to use the bricks as a tool. In detail, the purpose is (1) to introduce the objective of the workshop, (2) to introduce the workshop participants to each other, (3) to present team or stakeholder workshops as a tool and (4) to create an open-minded and 'safe' atmosphere, where everybody can feel free to share their insights and experiences.

Fig. 8.2 Exercise: Build a little house in 3 min of 10 random bricks

The first thing that happens in a team or stakeholder workshop is that the facilitator introduces the workshop objective. This introduction is made very simple—and in short sentences. The same goes for any presentation of the project or project scope, if there is a need for this. The facilitator tries to avoid any indirect suggestions of one 'right' perspective on the project or situation.

Next, the facilitator introduces the participants to each other. This is especially relevant in the stakeholder workshop where the participants often do not know each other. One of the easiest ways to introduce the participants to each other is by having them present their name, their job/interests and their reason for participating in the workshop. The facilitator notes the name of the participants and uses these in the workshop.

If the participants know each other, the facilitator can introduce a small exercise to let participants share information about themselves, which they probably have not shared with their 'colleagues' before. An example of an exercise like that is reviewed the following. The facilitator provides each participant with 10 random bricks, and asks them to build a little house in 3 min (see Fig. 8.2). When they have built the house, he asks them to rebuild it in way that tells something about their own house, their family, or their interests. The participants are given 5 min to rebuild the model, and afterwards the facilitator asks the participants one by one to explain their model to the rest of the participants.

8.3.1.1 Introduce Workshop as a Tool

With the introduction to the workshop objective and the introduction of the participants accomplished, it is time to introduce team or stakeholder workshops as a tool.

The introduction to team or stakeholder workshops are mainly done through small, humorous exercises, which show the qualities of it. However, some facilitators prefer to start the introduction with some background information. Others may find it sufficient to briefly introduce team or stakeholder workshops as a means to bring forward tacit knowledge and to create a shared language between the participants.

After the more or less detailed introduction to the background and qualities of team or stakeholder workshops, the small exercises are rolled out. It is common to have two or three exercises, which can be selected and further developed from the examples the following two sections.

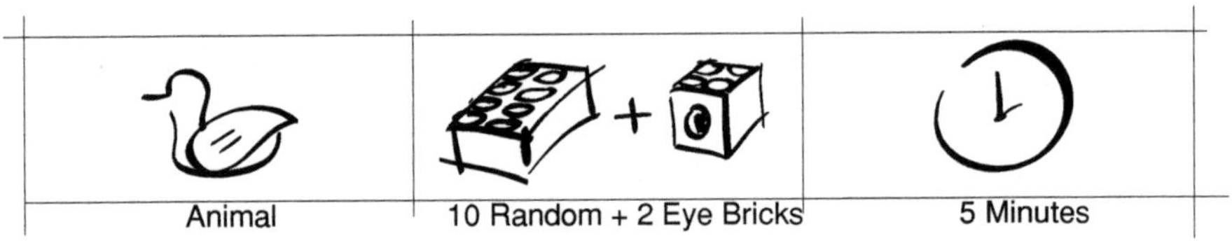

Fig. 8.3 Exercise: Build an animal in 5 min using 10 random bricks and 2 "eye" bricks

Fig. 8.4 Exercise: Build a long, tall bridge from 10 random bricks in 3 min

8.3.1.2 The Little Animal

In this exercise, every participant is given 2 small bricks with eyes, and 10 random bricks from which they have to build a little animal in just 5 min (see Fig. 8.3).

Having built the little animal, the participants are asked to explain this animal as a picture of a marriage or a manager.

Every participant explains their model (often followed by some laughter) and the facilitator ends the exercise with some concluding remarks—for instance:

> This exercise shows how many icons, pictures and meanings, we can imagine and attach to some very simple figures. It is therefore a good idea to pay attention to the details of the bricks and the connections between the bricks, when you build and look for the inspiration they hold.

8.3.1.3 The Bridge

In this exercise, participants are asked to select 10 random bricks. When the bricks are selected, they are asked to build the longest and tallest bridge possible in just 3 min (see Fig. 8.4). When the bridges are done, the facilitator says that he will now test if the bridges can make the mobile phone test. He walks around the table to the different bridges and places his mobile phone on it—just hard enough to make all the bridges collapse. When all the bridges are broken, the facilitator asks the following questions:

> What did you feel when I crashed your bridges? Did you feel that it belonged to you?

The questions can be followed up with some concluding remarks, for instance:

> This is another thing a team or stakeholder workshop is good at: creating ownership, when we build things. We become attached to them. And it is good that we become attached to

Fig. 8.5 Exercise: Build a model that shows what a good friend is like in 10 min

> the understandings, we create about this project. But it also means that you should be careful, if you alter someone else's model. Remember that they may be attached to it, and may want to take part in the alterations or at least approve the alterations.

8.3.1.4 The Friend

In this exercise, participants are asked to build a model, which shows what a good friend is like. They can use all the bricks they like, and they are given 10 min to complete the assignment (see Fig. 8.5). When the models are created, the participants are asked to explain the model made by the person sitting next to them. After each explanation, the participant who created the model is asked to comment on whether the other participant's explanation was close to what he/she imagined, when building it.

After every model has been interpreted and explained, the facilitator ends the exercise with some concluding remarks, for instance:

> 'The exercise shows how we can build meaning into a model and how we can interpret things from what is built. This is also useful in the next couple of assignments—if you think you are stuck: stop thinking and start building. The exercise also shows how many views we have on the concept: Friend, and it gives us a nuanced view of what a friend is and can be.'

Asking adults to build with bricks can be a bit of a challenge and especially the small exercises in the beginning are crucial in terms of convincing the participants to use the tool.

Throughout the introduction, the facilitator encourages the participants to take an interest in the assignments, while assuring them that they are on the right track—if they doubt it. Furthermore, if one of the participants misinterprets an assignment, and builds something different from what he/she is asked to do, the facilitator focuses on understanding what has been built, instead of correcting the participant. This is to show that there is room for mistakes.

8.3.1.5 Creating an Open-Minded and 'Safe' Atmosphere

The fourth purpose of the skills building part (to create an open-minded and 'safe' atmosphere, where everybody would feel free to share their insights and experiences) is probably the hardest and most important purpose to achieve.

It is partly achieved through the introductions and small, humorous exercises and partly through the facilitator's open-minded and non-judgmental attitude.

If it works successfully, some of the participants will start sharing personal insights and experiences even in the first few exercises, which opens up for other potentially hidden or tabooed personal insights to be shared later in the workshop.

8.3.2 Part 2: Individual Modeling

Whereas the purpose in part 1: Skills Building is to set the stage for the workshop, part 2 dives directly into the purpose of the workshop. It can therefore be useful to prepare participants for this by stating that focus will now be on the objective of the workshop and also assuring them that there will be no more surprises or humorous features.

The individual modeling is very much like the small skills building exercises. The participants are given an assignment, which they respond to by building a model. The model is presented to the other participants and there is time for questions and comments. The individual modeling is based on the workshop questions, which are rephrased into a set of assignments. For instance, if the question is:

> What is the X-project about?

The assignments will be:

> Your assignment is now to build a model, which shows what the X-project is about (seen from your point of view)

The assignment is presented verbally and participants are given between 20 and 35 min to build the assignment, depending on the complexity of the project and the question.

The participants are allowed to use all the bricks they want and are given half-time notice as well as a notice, when there is 5 min left. If a large part of the participants are not ready within the time limit, the building time is extended.

When everyone has finished, the facilitator asks the participant who finished first to start explaining his or her model. The facilitator asks detailed questions to help the participant unfold the meaning of the model. At the end of every explanation, the facilitator asks the other participants, if they have any questions or comments. The presentation of the models continues until everyone has presented.

If the plan is to have more than one assignment in the individual modeling part, it is important that the last assignment is the one in which there is a need for a shared understanding (The models, which are built for the last individual assignment, are the ones that are built together in the next part of the workshop).

8.3.2.1 Facilitation and Interventions

Like in the first part of the workshop, the facilitator maintains equal focus on all the participants and the insights they bring to the table. Some participants will explain for a long time, whereas others will give very short explanations. For those who make the short explanation, the facilitator follows up with some elaborating questions about their models.

During the presentations, the facilitator involves all the participants in the models being presented, by asking for their comments or potential questions, or by asking more directly:

> Do you recognize the perspective presented in the last model? and What are your personal experiences in relation to this?

8.3.2.2 Personal Experiential Concepts

In part 2, another important task for the facilitator is to look out for Personal Experiential Concepts, and draw the group's attention to these. This means that if any models have the characteristics of a Personal Experiential Concept, the facilitator adds focus to these by referring to their name in some of the follow-up questions or by asking the other participants, if they have experiences which are in line with the models (again called by their name/concept).

Another way for the facilitator to work with the Personal Experiential Concepts is to promote their development, either by asking for personal experiences, when a concept is presented, or by suggesting a concept based on the experiences presented. For instance, if there is a potential concept in one of the models like 'The Gate to the Elderly Life', the facilitator asks the participant, who created it, about its origin and the experiences behind it.

The aim is to get both the concept component and the experience component from the participant, if possible.

Alternatively, the facilitator suggests a concept, which can capture a number of experiences shared by the participants. Although this is a bit more difficult, it can be done, for instance by looking at the models and interpreting these—like it was done in workshop 2 (Red Cross/Base camp), where the facilitator interprets a model:

> Poul (facilitator): (…) Well… one of the things I find interesting here is the modular-structure, which you have created now. Is that a way of thinking, which you usually use? (…)

However, it is a very delicate balance and the facilitator is very cautious about not stepping out of his/her neutral role and taking ownership of the models away from the participants.

8.3.3 Part 3: Building Together

When the last set of individual models has been presented, time has come for the participants to build their models together. The building together process is presented as an assignment to the participants—for instance like this:

> Your assignment is now to build the individual models together into a shared model. You are free to select from all the individual models, and you do not have to use all of them. You can even choose to use a small part of an individual model.
> You are also allowed to build some new additions or connections, if you need it.
> The only rule is that: All of you have to agree on it.
> You have 45 min.

The process of building together can take anything from 45 min to 1½ h. Like in the individual assignments, the participants are given a notice when half of the time has gone, and when there is 10 min left. The time limit is not managed very strictly, and the facilitator can decide to give participants more time if needed to finish the assignment.

However, keeping a time limit helps to push the process.

8.3.3.1 Facilitation and Interventions

The building together assignment is often a hard nut to crack for the group of participants, and initially it may create a stop in the workflow. Sometimes it makes the atmosphere seem a bit tense; there can be long periods of silence and even some expressions of frustrations. The facilitator's role in this part of the process is very low key. He is mainly observing and only interrupting, if the process is no longer democratic or off-topic. However, the facilitator remains helpful and encouraging, while showing confidence in the group's ability to build the shared model. As described earlier, the process of building together often starts when one of the participants takes ownership in another participant's model, by suggesting that this should be part of the shared model. The facilitator can also help this process along, for instance by asking a participant:

> Which model do you think should be part of the shared model? Or which models do you find interesting?

Just to start the discussion. Even so, the facilitator does not at any point take responsibility for the assignment to create the shared model. His focus should remain on the process and the interactions. His job is to assure a fair and equal decision-making process.

When the building together process has been started, there are also some potential pitfalls the facilitator needs to look out for—and interfere with, if necessary.

One of the things that can happen in the building together process is that one or two participants take over control of the group. In this case, the facilitator maintains the democratic process by making sure that all the participants support every decision.

Likewise, if a participant chooses not to take part in the decision-making process, the facilitator interrupts the process, and asks what he/she thinks about the present decision. Another potential pitfall is if one of the participants starts systematizing or drawing conclusions away from the table of models—for instance on a blackboard. In this case, the facilitator asks the participants to work with the models on the table, and thereby putting the participants' focus back on all the models (perspectives) on the table—and not just the few selected on the blackboard. If the process slows down after some time of building together, a break can be a good solution.

8.3.3.2 Shared Experiential Concepts

Like in the individual modeling part, the facilitator looks out for Shared Experiential Concepts, and directs the group's awareness to these in the building together process, for instance by referring to them. However, it is not as elaborate as in the individual modeling process.

In the building together process, the facilitator can also note down when a Shared Experiential Concept appears and use these names in the workshop reflection as well as in documentation.

8.3.3.3 Reflection and Documentation

When the shared model is created, a short recap of the workshop and a reflection on what has happened should follow.

To recap the workshop, it is recommended to let one or two of the participants explain the shared model. This is also done to confirm that the participants agree on what has been built. It is further recommended that the explanation of the shared model is videotaped and that a number of pictures are taken of it. This is a very easy and valuable way of documenting the outcome of the workshop.

After the recap the time for reflections comes on the workshop as well as on the outcome of it. This can be structured in many ways, but asking one question at a time and letting everyone around the table answer it is probably the easiest one. The questions could for instance be:

What did you find the most interesting in the workshop today?
What do you think about the outcome?
What do you think about the workshop process?
Etc.

8.3.4 Facilitation Plan

One of the easiest ways to plan and execute a team or stakeholder workshop as a facilitator is by making a facilitation plan.

The facilitation plan is a guiding tool made by the facilitator and for his/her personal use only. It is like a screenplay, which shows the different things that have to happen in the workshop. It holds a time schedule, questions, and reminders for the facilitator.

A facilitation plan is a personal tool and can be very different from one facilitator to another—some like to have many details, while others do not.

The facilitation plan shown below is therefore to be seen as an example, which can be modified according to the needs and wishes of the facilitator. The facilitation plan is from workshop 4.

8.3.4.1 Facilitation Plan: Medical Treatment House

The facilitation plan example is given in the Table 8.1.

8.4 Results: The Outcome of a Team or Stakeholder Workshop

After the workshop, one of the main tasks for the project manager is to integrate the insights and shared understandings from the workshop into the project, using the visual material of the Personal and Shared Experiential Concepts as the main documentation and drivers to integrate the shared understanding from the workshop into the early phase project.

8.4.1 Future Reference Points

The very nature of the team or stakeholder workshop generates a lot of visual material that can be documented. The significant models that represent Personal and Shared Experiential Concepts should be photographed, and the participants presenting and explaining the models should be documented by video. These significant models of Personal and Shared Experiential Concepts represent important points of reference in the understanding of how the workshop participants answered the assignment questions. They are metaphors and symbols that encapsulate insights, storytelling, and second-order understandings from the discussions and negotiations of the participants in the team or stakeholder workshop.

They have an important role as a future reference point in any discussion and dialog concerning the project between the participants in the workshop, both as a stakeholder point of reference and as an internal point of reference for a development team. In short, this personal and shared meaning making could become part of the project 'language'.

Table 8.6 Facilitation plan example

Remember:
Selected bricks (18 with eyes)
Nametags
Camera/video
Watch

SECTION 1: PRESENTATION
9.00–9:15

- Introduction to the workshop aim and participants

  Aim

  To create a shared picture of a medical treatment house and to create guidelines on the basis of this

  Participants

  Name, job/interest, Reason for participating?

9:15–9:30

- Introduction to team or stakeholder workshops (Why using Lego Bricks?)

  Hand–mind connection

  Tacit knowledge

  Shared language

  Normally we use different languages: words, drawings, calculations, instructions etc.

  The process

  Assignment, build, present to other participants

SECTION 2: INTRODUCTION EXERCISES
9:30–11.00

1. The little Animal
2. Small bricks with eyes + 10 random bricks

   "Your assignment is now to build a little animal—you have 5 min."

   "Now please describe this little animal as a picture of a supervisor or a marriage."

   Concluding remarks:

   "This exercise shows how many icons, pictures and meanings, we can imagine and attach to some very simple figures. It is therefore a good idea to pay attention to the details of the bricks and the connections between the bricks, when you build and look for the inspiration they hold. Please save the model, we will use it again later on."

2. The bridge

   10 random bricks.

   "Your assignment is now to build the longest and tallest bridge possible with the bricks you have. You have 3 min."

   The mobile-phone test.

   Concluding remarks:

   "What did you feel when I crashed your bridges?"

   "Did you feel that it belonged to you?"

   "This is another thing a team or stakeholder workshop is good at: creating ownership, when we build things. We become attached to them. And it is good that we become attached to the understandings, we create about this project. But it also means that you should be careful, if you alter someone else's model. Remember that they may be attached to it, and may want to take part in the alterations or at least approve the alterations."

(continued)

Table 8.6 (continued)

10 min break

3. The rebuild assignment

The model from assignment 1 + all the bricks you may like

"Your assignment is now to rebuild your first figure, in a way which tells something about how you see/understand a medical treatment house."

"Please explain the model made by the person sitting next to you"

Creator comments on his/her model.

Concluding remarks:

"The exercise shows how we can build meaning into a model and how we can interpret things from what is built. This is also useful in the next couple of assignments – if you think you are stuck: 'stop thinking and start building'."

"Besides, the exercise shows how many views we have on the concept: Medical Treatment House, and it gives us a nuanced view of what it is and can be."

SECTION 3: INDIVIDUAL MODELLING

11.00–12.00

4. Individual modelling of Typical Problems

"We have now worked our way into the more serious part of the workshop—no more surprises. From now on you can also use all the bricks you want."

"Your assignment is now to build a model that shows some of the typical problems a patient/citizen may experience in relation to the present healthcare offer."

"You have 20 min"

Concluding remarks:

"Questions? Comments?"

"Is there any convergence between the models?"

"Is there any overlapping?"

"Is there anything we have forgotten?"

12.00–13.00 Lunch break

13.00–14.00

5. Individual modelling of Ideal Situation

"Your assignment is now to build a model which shows what a future medical treatment house must be like, in order to be of interest to you."

"You have 20 min"

Concluding remarks:

"Questions? Comments?"

"Is there any convergence between the models?"

"Is there any overlapping?"

"Is there anything we have forgotten?"

10 min break

SECTION 4: SHARED MODELLING

14.00–15.00

6. Shared modelling; Building together

"Your assignment is now to build the individual models together into a shared model.

(continued)

Table 8.6 (continued)

You are free to select from all the individual models, and you do not have to use all of them. You can even choose to use a small part of an individual model.

You are also allowed to build some new additions or connections, if you need it."

"The only rule is that: All of you have to agree on it."

"You have 45 min."

10 min break

15.00–15.30

7. Documentation and Reflection

Select one or two participants to explain the shared model and video-document it.

Reflection:

"What did you find most interesting in the workshop today?"

"What do you think about the outcome?"

"What do you think about the workshop process?"

8.4.2 Ownership of Insights

The project managers' participation in the workshop ensures that insight, experiences, and the value of the Shared Experiential Concepts can be carried directly into the management of the project without any need for translation, communication or extensive detailed descriptive documentation.

The team or stakeholder workshop in itself has no direct influence on the project; it is of course up to the management of the project (and company) to decide how to proceed. But the team or stakeholder workshop does offer the stakeholders, team members, and management a direct insight into each other's perspectives through a very tangible meaning making negotiation process, which can provide solid reference points for understanding the answers to the relatively abstract questions put forward during a workshop.

MIX
Papier aus verantwortungsvollen Quellen
Paper from responsible sources
FSC® C105338

If you have any concerns about our products,
you can contact us on
ProductSafety@springernature.com

In case Publisher is established outside the EU,
the EU authorized representative is:
Springer Nature Customer Service Center GmbH
Europaplatz 3, 69115 Heidelberg, Germany

Printed by Libri Plureos GmbH
in Hamburg, Germany